GEOGRAPHY

Towards a General Spatial
Systems Approach

GEOGRAPHY

TOWARDS A GENERAL SPATIAL
SYSTEMS APPROACH

William J. Coffey

METHUEN

London and New York

First published in 1981 by
Methuen & Co. Ltd
11 New Fetter Lane, London EC4P 4EE
Published in the USA by
Methuen & Co.
in association with Methuen, Inc.
733 Third Avenue, New York, NY 10017

Printed in the United States of America

British Library Cataloguing in Publication Data

Coffey, William J.
   Geography: towards a general spatial systems approach.
   1.   Geography—Methodology
   I.   Title
   910   G70
   ISBN 0-416-30970-4
   ISBN 0-416-30980-1 Pbk (University paperback 743)

To my parents
and
my little friend Janette

Contents

Contents

List of Tables

List of Figures

Acknowledgements

The author would like to thank the following editors, publishers, organizations, and individuals for permission to reproduce figures, tables, and quotations.

Editors

Bulletin of the Geological Society of America for 4.12; *Ekistics* for one quotation; *General Systems* for Table 4.1; Michigan Inter-University Community of Mathematical Geographers Discussion Paper Series for Table 4.2; *Papers and Proceedings of the Regional Science Association* for 6.7; *Scientific American* for one quotation; *Transactions of the Institute of British Geographers* for 6.8.

Publishers

Elsevier North Holland, Inc., for one quotation; Cambridge University Press for 5.5; Dover Publications for Table 5.1; Edward Arnold for 5.9; W.H. Freeman for 4.13; C.W.K. Gleerup for 2.1, 4.3, 4.6, 5.10, 5.11, 5.18 and one quotation; John Wiley & Sons, Ltd., London for 5.2; John Wiley Sons, Inc, New York for 6.11 and 6.12; The Macmillan Co. for one quotation; McGraw-Hill for one quotation; Prentice-Hall, Inc. for 4.23, 6.3, 6.6, 6.13, 8.4.

Organizations

Department of Geography, Atkinson College, York University for 4.19; Department of Geography, Northwestern University for 4.20.

Individuals

M.J. De Smith for 4.14, 4.15, and Table 4.3; William Warntz for 4.4, 4.5, 4.8, and two quotations; Martin Gardner for one quotation.

Preface

Three major themes have provided the impetus for this wide-ranging discussion of the discipline of geography: the 'systems approach', in its varied forms; the notion of a methodological unity to the discipline; and what might be termed the 'geometric-analytic' approach or the spatial viewpoint.

Ever since its explicit introduction into the geographic literature in the early 1960s, the notion of a systems approach has been largely perceived as having a modest degree of utility and a high degree of ambiguity. First came the exhortations. These neglected to specify the nature of the approach but assured us that it was the appropriate methodology for a scientific geography. A series of conceptual statements then underscored the need to recognize the interrelatedness of the various systems that exist on the earth's surface. The primary concern of geography was identified as the man-environment system, and the systems approach was seen as furnishing geography with a language common to other branches of science. This phase was followed by more substantive attempts to identify the systemic properties of real world objects and events and to utilize these properties in formulating and solving research problems. More recently, the systems approach has been employed in a pedagogic context. Many courses and texts now introduce the fundamental systems concepts, and rare is the undergraduate who has not been exposed to the basics of systems thinking. A major concern of the present volume is to attempt to lessen the degree of ambiguity surrounding the systems approach in geography and to demonstrate the conceptual and operational utility of the approach.

The notion of methodological unity within the discipline of geography is really a subject of broader epistemological (relating to the nature and bases of knowledge) and ontological (relating to the nature of being) questions concerning the unity of science and the unity of nature itself. The discussion of these broader questions dates back to classical antiquity and continues in the present era. In the contemporary period the view of a discernible unity to both science and nature has been advanced most explicitly by the advocates of General Systems Theory. Geography, like other fields of knowledge, does not exist in isolation. Thus, the debate over the unity of science and of nature has manifested itself both in periods in which the unity of the discipline has been striven for and/or recognized and those in which greater attention has been devoted to the individual, specialized sub-fields. The parallel aims of a diverse group of geographers who, during the late 1960s and early 1970s, sought to demonstrate the methodological unity of the discipline and those of the General Systems Theorists led William Warntz to describe the activity of these geographic researchers as 'General Spatial Systems Theory'. A second concern of this book is the development and the evaluation of the notion of a general spatial systems approach to the discipline and, further, the utility of this approach in the development of a methodologically unified field of geography and the role of such a unified geography as a spatial science integrating diverse areas of knowledge.

The 'geometric-analytic' or spatial viewpoint, with its emphasis upon the morphology of human and physical phenomena and upon the search for spatial theory, was considered in the 1960s and early 1970s to represent one of the foundations of modern geography. Today, this tradition appears to have assumed a secondary importance, having been largely displaced by an emphasis upon inductive statistics and upon dynamic models. There can be little doubt that the decline of the spatial tradition is associated with the recent waning of interest in the notion of unity within the discipline for that unity is most apparent at a level of abstraction in which diverse phenomena are considered in terms of their spatial properties, independently of their non-spatial characteristics. A third concern of this volume is to demonstrate that the geometric-analytic approach, extended so as to include an explicit consideration of the movements which are both cause and effect of spatial structure, not only continues to possess considerable utility within the discipline but, moreover, forms one of the bases of the concept of spatial process. Spatial process, in turn, is identified as the basis of the general spatial systems approach.

In the recent geographic literature the level of interest in systems and systems approaches appears to have been increasing. This interest is manifest in the excellent volumes authored by Bennett and Chorley (19) and Chapman (64), among others. On the other hand, concern with the unity of the discipline and with the spatial tradition now appears to be somewhat diminished relative to a decade ago. Therefore, the possibility clearly exists that some readers may view this treatment of geography as anachronistic. It is my belief, however, that such a view is not justified. While the emphasis upon the spatial tradition and upon methodological unity may have decreased in relative terms the import of these concerns within the discipline continues unabated. In addition, there are signs that with the continuing development of systems research and systems approaches within geography the issue of unity and the lessons of the spatial viewpoint are once again assuming a central role.

This book is an essay, in the original sense of the word – an attempt, an initial tentative effort. The attempt is made not only to present in a coherent manner the three broad sets of issues noted above but, more generally, to develop and to evaluate the notion of a general spatial systems approach to geography. As this approach has not been previously dealt with in a formal manner, further refinements may suggest themselves subsequently. In addition to being tentative, this volume is also, of necessity, selective. The range of topics considered is extremely broad and certain areas have been developed more fully than others. It is hoped, however, that this discussion will both stimulate and facilitate further detailed intellectual explorations on the part of the reader.

The criticism and advice of William Warntz, Don Janelle, Jim Pooler, and Mike Goodchild have been invaluable in the preparation of this book. Even more importantly, these individuals have been good friends. Any errors or omissions in the text are, of course, the sole responsibility of the author. A number of other persons whose enthusiasm and ability have indirectly influenced this book include Allen Philbrick and Elaine Bjorklund, Roger White, Jack Sommer, C.A. Hooker, and William Bunge.

Maps and figures have been expertly produced by Derry Graves and his cartographic elves, Randy Getty and Debbie DesRivieres. The Institute of Public Affairs, Dalhousie University provided a hospitable environment and financial support for this undertaking. Sarah Baker, Debbie Bulmer, Margaret Dingley, Margaret Douglas, and Donna Gordon managed many of the details of the manuscript preparation. Finally, my friends Diane Shillington, Barbara Janelle, and Judy Meyrick provided logistic support, free meals, and good humor. All of the above have my sincere gratitude.

William J. Coffey
Institute of Public Affairs
Dalhousie University
Halifax, Nova Scotia

PART ONE

GEOGRAPHY, SCIENCE, AND SYSTEMS

1

The Context

The focus of this essay is a set of concepts that together have been regarded as an organizing framework or conceptual structure[1] within the discipline of geography: the systems approach. During the past twenty years a systems framework, in one or another of its varied forms, has been utilized to examine both the internal cohesion of the discipline (3; 147; 167) and the relationship of geography to the other sciences (4; 218; 296), as well as to conceptualize and to operationalize specific research methodologies (6; 68; 82; 197). This approach has further manifested itself in books and courses explicitly organized around systems and systemic properties and attributes. Harvey (147, p. 449) has provided an influential imprimatur to the systems framework on the basis of his observation that, from both a methodological and an empirical point of view, 'the concept of a system appears absolutely central for our understanding of explanation in geography.'

Its recent wide utilization and, as Harvey tells us, its central role notwithstanding, the systems framework has been marked by a high degree of ambiguity. Terms such as 'systems approach', 'systems thinking', 'systems analysis', and 'general systems theory' have been employed interchangeably by a number of writers both in the geographic literature and elsewhere.

The purpose of the present attempt is fourfold. First, it seeks to clarify some of the vagueness and confusion that have become associated with the use of systems concepts and terminology in geography. The 'systems approach' is actually comprised of a number of loosely related approaches, and the fundamental distinction between systems analysis and general systems theory needs to be explicitly articulated. Second, it seeks to review and to evaluate the diverse attempts to introduce systems frameworks into geography. Third, it seeks to amplify upon the implicit and explicit use of systems approaches in geography. Finally, it seeks to develop and to evaluate the notion proposed by Warntz (337) that it is both feasible and useful to regard geography in the context of general spatial systems theory. This latter aim is viewed as particularly important, as the general spatial systems approach has methodological and epistemological implications for issues concerning both the nature of the discipline and its relation to other sciences.

Chapter 1 attempts to develop the context in which one may consider a framework, or several frameworks, based upon systems concepts. The first section introduces the general spatial systems approach and identifies a concept central to its formulation – spatial process. The second section places the notion of a framework in an historical perspective within the discipline of geography; the third examines several methods utilized for the study of systems; and the final section reviews some issues relating to the definition and interpretation of science. The latter is of considerable importance if one accepts the suggestion (3; 147; 218) that geography is in some way related to, or a component of, science. The scope and methodology of the discipline of geography are examined in chapter 2.

I. Towards a General Spatial Systems Approach

A. INTRODUCTION

One useful method with which to examine the implications of a general spatial systems approach is simply to analyze the terminology employed. First, it is a *general* spatial systems approach. The basic proposition upon which this conceptualization is founded is that there are certain laws which govern all physical and human phenomena; that is to say, there is a unity to nature manifest in a manageable number of general principles. Thus, a nomothetic, law-seeking methodology is an appropriate one to employ in the study of the real world. The vehicle most suited to this type of inquiry is theory (see pp. 29–30, 45–51). The construction of theoretical models, in the mathematical-logical sense, is one methodology that is characteristic of this approach. It is to be noted that theoretical models relevant to certain general processes and phenomena of space may be found in many different disciplines. This underscores the unity of knowledge about spatial relations in the real world.

Second, the approach is one involving general *spatial* systems. Briefly, 'spatial' may be construed to include both space-occupying systems and those theoretical and/or philosophical systems which are in part responsible for the organization of space. In both cases, emphasis is upon both spatial form and movement in space. It is important to note that neither structure nor movement can be identified as exclusively causal or effectual, as form and movement are very tightly interrelated. This notion is further developed below as spatial process.

Third, the approach is one of general spatial *systems*. The construct 'system' and its manifestation in systems thinking are included because both have proved to be extremely useful in several respects. First, a system provides a way of treating the complexity that exists in the real world, especially a way of handling specific problems that may have complex aspects. Systems thinking often furnishes concepts that may begin to elucidate the function and structure of complex sets of relationships. Second, several theories, the Second Law of Thermodynamics for example, have been identified which are of fairly general applicability to systems of varying type and scale. These theories may hasten our understanding of the structure and movements underlying the spatial distribution of terrestrial phenomena. Third, the systems approach recognizes the complementarity of various modes of inquiry. This has special relevance in geography, for Anuchin (10, p. 258) has noted that the problem of the general and the particular that has divided geography is essentially the problem of the relationship between analysis and synthesis.

In sum, a general spatial systems approach offers a method of viewing the complexity of the real world and specific problems within it, the discipline itself, and our knowledge of nature. Before specifically considering the organization of a general spatial systems approach, let us explore in greater depth some of its ideal characteristics.

B. CHARACTERISTICS

The observations made in this section concerning the nature of a general spatial systems approach are in large measure normative. Since a formal general spatial systems framework has not previously been articulated, the present discussion concerns itself with what should be, rather than that which exists.

First, such a framework should be organized in such a way as to demonstrate the unity of nature, the unity of science, and the unity of the discipline. This general theme will be extensively dealt with in the remaining portion of this chapter and in chapter 2. For the present, suffice it to say that the position that order exists in nature, particularly in the spatial manifestations of natural (human and physical) and artificial (man-made) phenomena, is a valid one philosophically. Any science that attempts to discover and investigate this unity must possess an internal logical unity if it is to stand a chance of success.

Second, the framework must rest primarily upon deductive reasoning. The course of science has demonstrated that the deductive mode is associated with the majority of significant theoretical and empirical advances.[2] The inductive mode is, of course, an important and indispensible complement.

Third, the complementarity of approaches should be stressed; it must be recognized that there is no one exclusive answer to a problem, nor an exclusive perspective from which a problem must be viewed. The real world is highly complex and multi-faceted. Two distinctly different, even contradictory, approaches may each be valid and entirely worthy of pursuit. Complementarity manifests itself in the geographer's ability and willingness to use both inductive and deductive reasoning, and analytic and synthetic methodologies.

Fourth, the framework should be one that organizes and simplifies knowledge; it should produce economy of thought of the sort indicated by Georgescu-Roegen: 'a catalog which lists known propositions in a logical – as distinct from taxonomic or lexicographic – order' (113, p. 26). A small number of broadly applicable principles and organizing concepts are to be sought.

Fifth, a general spatial systems approach must meet a standard of clarity. Bunge (55, p. 2) argues that clarity is best achieved when the propositions or theories are stated formally by use of mathematics, since this ensures explicitness and freedom from contradiction. As Richardson (254, p. 481) states:

> The magnificent conception of mathematics as the study of all abstract logical systems or abstract mathematical sciences and their concrete interpretations of applications really justifies the statement that mathematics is basic to every subject forming part of the search for truth. In fact, mathematics, thus conceived, includes all subjects into which one injects logical structure.

To this Kemeny (168, p. 33) adds that 'all scientific theories – numerical or other – are mathematical. This fact rests on the nature of Mathematics, on its identity with advanced logic.'

This emphasis on clarity and on the use of mathematics to achieve it raises the broader issue of the tools appropriate to the study of general spatial systems. Theory, the combination of fact and logic, is the foundation of this approach. Logic is particularly important as, by definition, it is not concerned with specific examples or unique individuals; logic deals with the members of a class. Logical reasoning, however, may be extended to individuals as members of a class (348, p. 251). One particularly useful application of logic is set theory. Warntz (339) has demonstrated that the exercise of regionalizing in geography is relatable to the operations performed on sets in mathematics. 'Translated into the terms of real space upon the earth's surface, unions, intersects, and subsets are valid ideas related to "uniform" geographical regions' (348, p. 250). Thus, the earth's surface is regarded as the 'set of all sets', and Venn diagrams and Boolean algebra, the algebra of sets, are employed by Warntz to perform logical operations upon its sets. Further, the geographical map is conceptualized as an extended Venn diagram. The Venn diagram, which portrays the relations among non-spatial sets, has its spatial counterpart in the map, which indicates the geometrical, as opposed to simply topological, relations between various sets.[3]

In addition, calculus, probability, symbolic logic, the various algebras, topology and the various geometries may be employed to enhance the clarity and precision of this approach to geography. Mapping, in its mathematic sense of proceeding from domain to range according to some function, and group theory (55, p. 215), the branch of mathematics that examines coherent sets of mathematical operations and determines the properties that remain unchanged by the application of the operations, are similarly useful tools.

Further, an integral tool of the general spatial systems approach is one that is often overlooked because of its very simplicity – graphics. Warntz (348, p. 249) points out the interrelatedness of the three 'very old and distinguished disciplines of geography, geometry, and graphics' and notes that early ties between them arose through projective geometry and the ancients' regard for maps. Graphical methods have proved useful for yielding approximations of solutions in cases where mathematics has proved to be intractable. Further, graphical thought and methods can be utilized not only as a means of portraying results but also as a means of experimentation. The infinite variety of projections and their spatial transformations, which allow us to expand, collapse, dislocate, repeat, invert, eliminate, interrupt, or superimpose surfaces, make maps the instruments of an experimental science (55, p. 61). Thus, maps may be regarded not only as media of storage for spatially ordered information, but also as the graphical expressions of solutions, and as the laboratory of geography.

Sixth, the framework must strive to attain the highest level of explanation, which implies the ability to predict. The predictability of spatial phenomena, in turn, depends upon the answer to the question of whether these phenomena are unique or general (55, p. 7). The matter of generality versus uniqueness will be dealt with below (pp. 26–28). Although uniqueness is an inherent property of all objects, at an abstract level this uniqueness begins to disappear and commonalities which enable phenomena to be grouped into classes are able to be readily identified.

Finally, the general spatial systems approach ought to possess a set of explicit organizing concepts or themes in order to furnish geography with both a theoretical and an empirical focus. It is this matter that we consider next.

C. A GENERAL FRAMEWORK

Schaefer (264, p. 226) tells us that methodology proper deals with the position and scope of a field within the total system of sciences and with the character and nature of its concepts. Further, methodology thrives on change and evolution; in an active field concepts are continuously either refined or discarded in whole or in part. The present attempt may be viewed as methodological in this sense. A framework for the discipline of geography is posited, one that is based upon considerations of the position and scope of the discipline within the context of science.

1. Process

A major source of ambiguity within the discipline and, to some extent, a major point of contention has been the definition of process and its perceived role within geographic explanation. Since the concept of process occupies a central position in the framework that will be advanced in this section, it is necessary to consider in some detail the use and misuse of this term within geography.

The concern with process in science may be traced back to the process metaphysics of Heraclitus (25). Where Democritus viewed nature as composed of a set of objects in the void, Heraclitus argued that all things are in a state of flux and that Becoming is the essence of living and being. This was perhaps the earliest attempt to reconcile time flow with space.

7

The work of Einstein notwithstanding, a great deal of progress in reconciling time and space has not been made, for Harvey (147, p. 413) notes that within geography, the science of space, a basic methodological problem is still that of finding an acceptable interpretation of time.

Compounding the methodological problem of an acceptable interpretation of time is the lack of specificity with which the term process is employed within geography. A process may be defined as a continuous and regular succession of actions taking place or carried on in a definite manner and leading to the accomplishment of some result; a continuous operation or series of operations (228, p. 2311). With this definition of process as a succession of actions through time we may contrast the less rigorous usage that has come to serve as a general catch-all phrase for just about anything that may be occurring, or for any sort of undefined mechanism. Harvey (147, p. 419) makes a further distinction between the above 'ordinary' usages of process, in the sense of any sequence of events over time, and its usage in the context of explanation; a sequence of events over time that are connected by some mechanism. The distinction is that any sequence of events cannot necessarily be regarded as explaining anything in particular, whereas explanatory principles may be inferred from a sequence connected by some established mechanism. Even Harvey's scientific usage of process is not without ambiguity, however. Thus, in geography we encounter terms such as 'the process of movement of population' and 'agglomeration and decentralization processes', and we have no way of knowing whether the person using these phrases is referring to the sequence of events involved, the mechanism, or both.[4]

Within geography an especially important consideration is the validity of the term 'spatial process'. The definition of process examined above tells us that process involves change over time. The real question is, however, whether process involves change over time alone. Specifically, there is a position taken by some geographers that process implies only the temporal dimension and that the term 'spatial process' holds no real meaning. The antecedents of this view may be traced back to the Kantian notion that both space and time are absolutes and are separate from one another. Blaut (37, p. 2) points out, however, that the Kantian view has been generally superseded by a relativistic philosophy in which space and time are both inseparably fused. Thus, nothing in the real world is uniquely temporal or uniquely spatial; spatial structures in the real world are simply slow processes of long duration (37, p. 4).[5] 'Spatial process', therefore, may be seen to contain real meaning and may not be dismissed as a logical contradiction.

Schaefer's (264) discussion of geographic methodology and the nature of the discipline is based upon a view of space and time that is somewhat Kantian or, more correctly, Newtonian. After presenting a long argument against the 'exceptionalist' view of Hartshorne (145), that is, against the view that geography is methodologically unique, Schaefer identifies one major aspect in which geography differs from the other social sciences. The latter, as they mature, concentrate on process and the discovery of process laws. Given the state of a system at a certain point in time, process laws allow for the prediction of changes that will take place (264, p. 243). On the other hand, says Schaefer, geography is essentially morphological. Purely geographic laws contain no reference to time and change. 'This is not to deny that the spatial structures that we explore are, like all structures anywhere, the result of process. But the geographer, for the most part, deals with them as he finds them, ready made' (264, p. 243).

Illustrative of Schaefer's view is his allocation of the laws of geography to three categories. In the first are the laws of physical geography, many of which are not strictly geographical as they were discovered independently in the physical sciences. In the second category are the laws of human geography, called 'economic geography' by Schaefer in the Germanic tradition. 'As far as they are morphological these laws are genuinely geographic' (264, p. 248). To the extent that these laws are not entirely morphological they belong to the third category, that of process laws. Schaefer concludes that such laws are not purely geographical, but that neither do they belong entirely with any other discipline; spatial variables are part of these process laws but they are not self-sufficient. Thus, Schaefer may be viewed as an exceptionalist in his own right, establishing a form–process dichotomy within the discipline of geography. The validity of this dichotomization will be discussed in the final chapter.

Harvey takes an approach similar to that of Schaefer, stating that process laws deal primarily with change over time (147, pp. 80-1).[6] Within geography he identifies a dichotomy between indigenous morphometric postulates and derivative process postulates, suggesting that synthesis in the discipline can be interpreted as a matter of linking theory governing process, which is mainly derivative, to theories about spatial form (147, p. 127). Hagerstrand's diffusion theory (139), Losch's central place theory (190), and the fluvial studies of Leopold, Wolman, and Miller (186) are cited as models of interaction between process and spatial form.

Bunge (55) was heavily influenced by the work of Schaefer and attempted to resolve Schaefer's form–process dichotomy, first by interpreting Schaefer's exclusion of *process laws* from geography as not also excluding *spatial process laws*. 'He did object to process laws . . . as non-geographic, but these non-spatial processes are another matter' (55, p. 210). Then Bunge seeks to reconcile the form–process distinction by pointing out that instead of the static (form) and the dynamic (change) being contradictory concepts, they are dual expressions. 'The duals can be designated as *spatial process*, meaning movement over the earth's surface, and *spatial structure*, meaning the resulting arrangement of phenomena on the earth's surface – the distribution' (55, p. 211). Warntz (347, p. 181) speaks also of 'spatial structure and spatial process – those coequal duals of form and movement.' Thus, both Bunge and Warntz equate movement with spatial process; they make this identity explicit in the following passage from an unpublished manuscript that they coauthored, *Geography – The Innocent Science:*

Any geographic explanation or spatial prediction of the location of the phenomena on the earth's surface – human, physical, or whatever – involves reference to movement such as circulation, diffusion, interaction, orbits, and flows. This is what distinguishes geographic processes from other kinds of processes. These movements, whether of eroding rivers on the earth's mantle or the migration of people from rural to urban areas, produces changes in the arrangement of features – that is, changes in the geometry. Geometry is general enough to embrace all the sub-concepts of spatial structure . . . Geometry and movement, viewed in this manner are linked duals. One interacts with the other . . . . The dynamics of geometry and movement are not separable. These are the coequal approaches which together constitute spatial relations. (348, p. 9).

The position that is taken in the present discussion is that, although process undeniably involves a temporal dimension, to regard process solely as a temporal construct is to acknowledge the view of time and space as distinct and separable. This, it is argued, is not a particularly useful manner in which to conceptualize the real world. Further, the view that process involves time alone is incompatible with certain operational aspects of the study of process. For example, one general process that may be identified is growth. If some object or organism is growing in a positive manner, we can expect it to be larger by some degree at time T_1 or T_2 than it was at time T_0. In fact, if growth is occurring at a constant rate, we should be able to predict the size of the organism at time T_3, and we should be able to infer the size of the organism at time T_{-1}. Growth is a process and undeniably occurs through time. Certain approaches to the study of growth, however, have shown that it is not necessary to consider time explicitly in order to make predictions and inferences about the growth of an object or organism. The theory of allometry, one of the earliest postulates of theoretical biology (161), concerns itself with relative growth, the increase in size in one component of an organism relative to the increase in a second component, or relative to the increase in the entire organism. The allometric approach has been demonstrated to be a tremendously useful tool in the biological sciences (220; 363). Although allometry involves an examination of the growth process, the temporal dimension of growth is not treated explicitly. Rather, growth is treated as a *spatial process* with the continuous changes in the shape and form of an organism emphasized. Processes do occur in space and time, but are exclusively in the realm of neither. Although it is an essentially analytic enterprise, it is often convenient to focus operationally on either the temporal or spatial dimension of a process (Figure 1.1), but this is not to deny that process involves the complementary and philosophically inseparable constructs of time and space.

Recognizing that time and space are relative and inseparable, we find that process must involve continuous change over both time and space or, if you prefer, over *time-space*. In geography, then, it is valid to speak of spatial process. This makes explicit the implicit notion of change over

A. Growth-in-Time (Absolute Growth)
   t=time   n=number or size

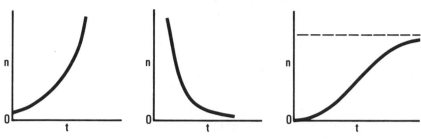

1. Positive exponential growth.   2. Negative exponential growth.   3. Logistic growth.

B. Relative Growth (Allometry)
x and y are two systems components

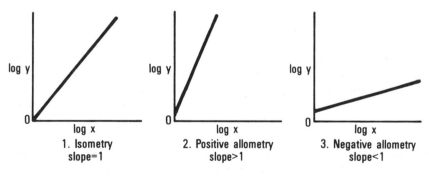

Figure 1.1 *Two dimensions of the growth process*

space and adds it to the already explicit view of change over time. In his discussion of the 'arrow of time,' Layzer (181, p. 68) even resorts to the use of the term 'temporal process' to make one of his points as explicit as possible. Similarly, Thom (307, p. 38) refers explicitly to 'spatial process' in his theoretical treatment of morphogenesis. Again, it is an operational convenience, particularly within geography, to view changes in form or structure apart from the implied time dimension. We may, in turn, resolve spatial process into two components, spatial structure or form, having no reference to change, and movement in space, which by definition must also occur through time.

Bunge and Warntz arrive at these two components also in their 'coequal duals of form and movement'. These authors take a very limited view of process, however, when they equate process with movement: 'The duals can be designated spatial process, meaning movement over the earth's surface, and spatial structure, meaning the resulting arrangement of phenomena on the earth's surface – the distributions' (55, p. 211). As an organizing framework, this conceptualization has one major drawback, it is divisive in that it establishes yet another dichotomy in the discipline of geography, in this case between form and movement. If one is willing to accept the argument made in chapter 2 for viewing geography as an integrating science that transcends the arbitrary compartmentalization of knowledge and, indeed, that integrates knowledge around the theme of space, the distinction between morphology and movement, even under the terminology of 'coequal duals', can only be viewed as contrary to the goals of coherence and synthesis.

Let us consider morphology as an intermediate and transitory state that is inexorably intertwined with changes over time and space in a spiral of cause and effect or, more exactly, circular and cumulative causation. Process, then may be viewed as a feedback loop in which morphology is not only the result of changes, but to a large extent plays a causal role by setting the initial conditions or constraints for the processes. There is no unique cause nor unique effect; all is interrelated through a complex series of feedback loops. In a sense, the relationship may be viewed as dialectical; everything is both cause and not cause, effect and not effect. Chisholm (66, p. 94) agrees with the arbitrariness of dichotomizing spatial structure and spatial process (movement), noting that the distinction is essentially a matter of the time period adopted or, in other words, of temporal scale. Abler, Adams, and Gould (2, pp. 60-1) expand upon this, pointing out that process and structure are, in essence, the same thing. Whether we see process or structure when we look at a spatial distribution depends on the time perspective that we adopt and the rapidity with which the process moves.

Although the distinction between the usage of Bunge and Warntz, both separately and collectively, of the terms spatial structure and spatial process and the schema introduced above (spatial structure and movement in space being synthesized to yield spatial process) may seem to some merely a matter of semantics, the discrimination is, in fact, a crucial one. The form–process dichotomy leads to analysis rather than synthesis; beyond a certain point (which has probably been reached already) it imposes limits on the ways in which we can conceptualize spatial relations. In addition, it is necessary to make distinctions over what may be called semantics, as the language that we employ constrains both the manner in which we think and the things we are able to think about (355).

2. Spatial Process: Structure and Movement

Spatial process, emphasizing the dynamic mutual causal feedback between structure and movement, is a strong candidate for the primary unifying concept of a general spatial systems approach. Neither structure nor movement is a sufficiently comprehensive concept for discussing the interrelatedness which characterizes the existence and the operation of a system. As is appropriate to the methodology and epistemology underlying the general spatial systems formulation, the concept of spatial process is a synthetic one, integrating spatial structure and movement in space (Figure 1.2). Further, as Figure 1.2 indicates, two related aspects of spatial process may be identified – growth and organization. These, in turn, are closely related to the principles of optimality, equilibrium and conservation.

In chapters 7 and 8 growth and organization will be examined as manifestations of spatial process which are applicable to the study of spatial systems having diverse scales and non-spatial referents. Before attempting to relate growth and organization to spatial process, however, it is necessary to develop an appropriate vocabulary and set of tools. This task occupies chapters 4 to 6, in which the elements of spatial structure and movement in space are examined. The manner in which these elements may be applied to the analysis of both human and physical phenomena will receive major emphasis.

Chapter 9, the concluding chapter, gives further consideration to the concept of spatial process. As noted above, the notion of process has always caused substantial difficulty within the discipline of geography. While spatial process is a formulation that has considerable utility for the analysis of spatial systems, some deficiencies remain to be overcome.

With this general introduction to the general spatial systems formulation and spatial process behind us, we may proceed to examine the broader context of this framework.

II. Frameworks in Geography

From an historical perspective, there have been a series of attempts to discover order in, or perhaps more correctly to impose order upon, the discipline. The theme of the internal coherence of geography has had, in very general terms, three major manifestations in the modern era; each of these has had associated with it a particular view of the relationship of geography to science. The first manifestation is identified by Wrigley (374) as the 'classical' approach. Both Humboldt and Ritter, critical of the descriptive and ill-organized treatments of geography on the part of their predecessors, advanced the view that the scientific organization of knowledge consisted of two parts: the careful collection of detailed and accurate factual material and the coherent organization of the material by the formulation of simple and concise cause and effect laws. The latter was viewed as the distinguishing feature of a science; without it any branch of learning was simply a taxonomic collection of facts. A related aspect of this 'classical' approach was the conviction, shared by such philosophers as Comte and Mill, that in the final analysis there was no methodological differentiation between the physical and social sciences. This brief account

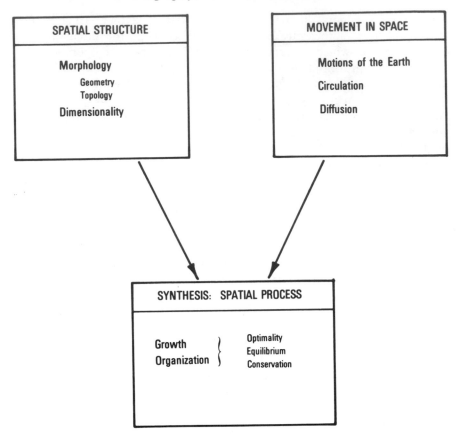

Figure 1.2 *A general spatial systems paradigm*

is, of course, a vast oversimplification of the positions of Humboldt and Ritter and the others, such as Ratzel, who succeeded them. It does convey the important point, however, that geography was conceptualized as a methodologically unified science possessing a theoretical organizing framework, the goal of which was the formulation of universal cause and effect laws. The decline of this approach as a consequence of the geographical determinism school of thought is too well known to require comment.

Whereas the classical approach to the organization of the discipline involved the view that a science was characterized by a particular methodology, the subsequent major approach, 'regional geography', was predicated on the notion that the primary criterion of a science was the subject matter which it examined. Further, whereas the classical view unified the human and physical aspects of the world within the context of the causal effect of environment upon man, the regional view held that human and physical objects and events were related through a phenomenological inseparability. That is to say, the influences of nature upon

14

man were not seen as distinguishable from those of man on nature. The area in which the intimate connection between man and nature had developed over time formed a unit, the region, which was the proper subject matter for the science of geography. And since each region was so much a product of local circumstances, both social and physical, what was significant in one area might prove to be irrelevant in another. Thus, a regional, rather than a 'systematic' or topical, treatment of the earth was a necessity; the region became the organizing concept of the discipline. As Fenneman (108) told us: 'the one thing that is first, last, and always geography and nothing else, is the study of areas in their compositeness or complexity, that is *regional geography.*' The specific variations on this theme include the work of Vidal de la Blache (326), Hartshorne (145), and Sauer (263).

The third and most recent organizing framework in the discipline may be identified as the spatial view. Largely supplanting the regional approach, the spatial view regards the role of terrestrial space as the primary organizing concept of geography. Although geographers have always been concerned with space, it was not until the emergence of a body of conceptual and empirical work which made explicit the influence of distance and relative location upon both social and physical objects and events (for example, that of Christaller (70), Schaefer (264), Bunge (55), Tobler (315), Haggett (140) and Ullman (323), to name just a few) that this approach became recognized as one of the foundations of contemporary geography.[7] Once again, geography has come to be regarded, at least by some researchers, as a law-seeking science; the definition of science adopted is one which emphasizes methods and conceptual tools over subject matter. Chapter 2 expands upon the spatial view, probably the dominant approach in contemporary geography.[8]

One manifestation of the spatial view is theoretical geography. Throughout much of its existence, including the present period, geography has been conceptualized in terms of a human portion and a physical portion, with many topical sub-specialties within each. Through the generality and abstraction of an increasing number of theoretical investigations, however, the discipline has begun to witness a trend toward reunification, as inquiry into the phenomena of general spatial patterns and processes has proceeded to a point at which basic isomorphisms between the human and the physical realms may be demonstrated. Indeed, one tenet of the theoretical–predictive movement in geography has been that the distinctions between the various 'systematic' branches of the discipline diminish at the theoretical level as common spatial properties and dimensional problems are recognized, despite their vast non-spatial dissimilarities (348, p. 225).

A problem not epistemologically unrelated to a consideration of the unity of human and physical phenomena is that of whether these phenomena are amenable to generalization. Logically, if they are not, the premise of the utility of generality and abstraction, upon which theoretical geography is so heavily founded, may be demonstrated to be faulty, and theoretical geography, itself, dismissed. The concern with generalization in geography is not a recent development, although its rise to prominence was precipitated by the Hartshorne-Schaefer dialogue. The debate still rages on in contemporary literature.[9]

Without doubt, the uniqueness of every small part of the earth's surface is there to be seen, but general regularities and systematic relationships may be and have been discovered. Theoretical geography, which seeks to articulate these general relationships, deals, as its name implies, not with fact but with theory – that which is a mixture of fact and logic (348, p. 9). Individual facts, the accumulation of which is the essence of the regional technique of complete spatial accounting, are of importance only in their relationships to other facts and as they contribute to the broad patterns that may be observed and meaningfully interpreted (334, chapter 4). Dolphin (92, p. 88) observes that in geography the hope is at once to avoid the 'bogey' of unique infinite particulars, and to regard the facts of areal arrangements as 'general' or 'generic', rendering them thereby susceptible to generalization and theoretical elaboration. Further, with the loss of the ability to construct general propositions comes the loss of explanation and prediction, for these are generated by deduction from statements on a general level.

III. The Study of Systems

The concept of a system is not a recent one; for many years, predating even the birth of Christ, scientists and philosophers have written of celestial systems, circulatory systems, and political, economic and social systems. The more rigorous emphasis upon formally defined systems that has arisen in recent years, however, is the result of a recognition of the system as a construct which is necessary to represent the complex set of interrelationships that exist in the real world.

As conventionally employed in science, the term system refers to a set of objects together with relationships between the objects and between their attributes (143). This definition may be illustrated by considering a system of cities. Here, the objects are the cities, however they may be defined; the attributes of a city might include the size and composition of its population or its economic characteristics; the relationships between cities would likely involve the various types of flows between them; the interrelations between attributes might involve qualitative or quantitative aspects of the relation between the sizes of the populations of two or more cities, as in the gravity or potential model. The systems conceptualization stresses the dynamic interrelatedness of objects in an holistic manner; emphasis is upon the relationships of the elements that compose the system rather than on the specific substances or characteristics of its elements. It is the relationships between objects and between the attributes of objects that often enable the researcher to define a system. More detailed reviews of the definition and properties of systems may be found in Harvey (147, chapter 23) and Hall and Fagen (143).

A variety of approaches may be taken to the study of systems. This diversity is reflected in the manifold terminology that has been noted above. There is, further, a marked lack of consistency in the manner in which these terms are employed. What one writer labels as systems analysis, for example, may be regarded as general systems theory by a second. Since a central concern of this discussion is the utilization of systems frameworks in the discipline of geography, it is crucial that the distinguishing characteristics of these approaches be identified and that a consistent set of terminology be established for discussing them. This is no mean task, in that not only does each approach share certain characteristics with its alternatives, but there are also numerous sets of criteria available for distinguishing between the approaches.

Bertalanffy (28, pp. xix-xxii) has identified three aspects of the study of systems. First, there is systems science, which deals with the scientific exploration of systems and the theory of systems in the various sciences. Second, there is systems technology, which deals with applications in both computer operations and theoretical developments such as game theory. Finally, there exists systems philosophy, which involves the reorientation of thought and world view ensuing from the introduction of 'system' as a new scientific paradigm. Systems philosophy is divided by Bertalanffy into systems ontology, which concerns itself with the meaning of 'system' and the realization of systems at various levels of observation; and systems epistemology, which undertakes the investigation of organized wholes of many variables. Systems philosophy seeks the integrative links underlying not only systems but also different realms of knowledge; it seeks to discover the interconnections between independently researched and formulated theories and, more importantly, seeks to ask those questions that most need to be asked in order for knowledge of the world to be advanced. Laszlo (179, pp. 4-10) speaks of systems philosophy as an endeavor which is able to provide a world view, which in the history of science has proven to be most significant for asking the right questions and perceiving the relevant states of affairs.

Bertalanffy's classification reflects the primary distinctions among systems approaches that are implicit in the contemporary literature. Since the language generally employed throughout the literature is more widely accepted than that utilized by Bertalanffy, alternative approaches to the study of systems will be delineated using the former.

A. SYSTEMS ANALYSIS

In contrast to other approaches, systems analysis is characterized by a *specific* set of purposes and analytical techniques. Although recognizing that all objects have connections with many other objects and that the significance of any one depends upon its interrelations with the others (208), the *analysis* of a defined system is the primary feature of the

17

approach. Therefore, complex and interrelated wholes are resolved into more elementary units for the purpose of investigating the response of a system to a given input or a given set of environmental conditions. Causal connections between input and output are sought in order that the effects of a change in some input variable or system element may be assessed. Thus, the *functioning* of a system under a variety of conditions is of paramount importance.

Systems analysis is generally regarded as the domain of the practitioner rather than the researcher. Its goals involve the control or optimization (or, at very least, the accurate prediction) of the behavior of a system, and the identification of the most feasible method for accomplishing a given purpose (358, p. 83). Systems analysis is aimed, therefore, at understanding rather than at explanation. One practical application of the approach is to expedite logical and coherent decision-making (71) with regard to the design and operation of a system. Thus, the methods used are mainly normative, often including some attempt to maximize a critical function (131, p. 27).

In order to *operationalize* the above goals, systems analysis seeks the development of the necessary sorts of technical apparatus. The creation of very specific and highly rigorous models is a major emphasis, as are such areas as communications engineering, cybernetics, information theory, and operations research. Steinitz and Rogers (286) even offer the view that systems analysis and operations research are synonymous. This is probably an overly limited view of the approach, however.

A useful partial summary of the systems analysis approach is presented by Kaplan (165):

(1) The system to be investigated is explicitly distinguished from its environment.
(2) The internal elements of the system are explicitly stated.
(3) There are relationships between the system and its environment that are explicitly stated.
(4) Where these relationships involve deductions the canons of logical or of mathematical reasoning are employed.
(5) Assertions concerning the relationships between the system and the real world are confirmed according to the canons of the scientific method.

As he is employing the term systems analysis in a slightly different fashion, Kaplan does not explicitly recognize the operational and performance oriented aspects that have been identified here as central to the approach.

B. SYSTEMS THEORY

Within the set of systems approaches, the distinction between systems analysis and 'not systems analysis' is quite readily identified. It is more difficult, however, to establish a basis for discriminating between systems theory and General Systems Theory (GST).

Systems theory, or general systems theory as it is sometimes known, is essentially a derivation of classical GST. Many of the focal concepts and methods of systems theory were first established in the classical version where, due to a different emphasis, they have been somewhat secondary to the overall theme of homology among systems of various types. Nevertheless, the influence of Bertalanffy's GST (28) is obvious in systems theory.

Systems theory deals with the properties, types, and behavior of systems on a more general level than does systems analysis. While the specific properties and behavior of defined systems are of interest, of more central concern is the concept of a general system, one that subsumes the characteristics of various real systems and allows one to discuss problems within a unified analytical framework. A general system is, then, a kind of higher-order generalization about the variety of real systems which individual disciplines have identified (147, p. 474).

Systems theory is typified by the analytic–deductive attempts of Mesarovic (210) and of Ashby (13) as a 'theory of general systems.' This involves not merely seeking isomorphisms in natural systems and generalizing their common characteristics into a broad set of principles but, further, the establishment of some general theory from which the characteristics of various systems can be deduced. Mesarovic, especially, is concerned with the establishment of a unified framework for discussing the various syntactical structures used for modelling real systems. His concern is more with a theory of general models, as opposed to the general theory about reality which is sought by GST.

It is the methodology, as distinct from the philosophy, of the study of general systems which characterizes systems theory. The substantive aspects of the approach include the consideration of the structure, behavior, elements, relations, state, boundaries, and environment of systems *in general*. One facet of this is the creation of typologies of systems. For example, systems may be classified as adaptive, controlled, dynamic, self-regulatory, and so forth. In addition, concepts such as dynamic equilibrium, equifinality, homeostasis, information, systemic hierarchies, feedback, entropy, negentropy, and closure (open versus closed boundaries) comprise the foundation of systems theory.[10] Such concepts also form the basis of GST, but are employed there in a less rigorous manner.

The systems theory approach is at a higher level of abstraction than systems analysis, as it is concerned with the general properties of diverse classes of systems. The major criticisms that have been levelled at it involve the difficulty of correctly determining the fundamental terms and theories that should be applied to systems in general, and its lack of operational utility relative to systems analysis. Systems theory is, in turn, a good deal more specific than General Systems Theory, which may be viewed as the highest level of abstraction in the study of systems.

C. GENERAL SYSTEMS THEORY

General Systems Theory is essentially a statement of a synthetic philosophy. While the notion of a general system is important in itself and in its application to the modelling of broad classes of phenomena, its role in GST is as a tool in the attempt to unify knowledge about the real world. GST is an attempt to put back together again those pieces of reality that science has 'dismembered for the purpose of analysis'; it is a general theory not just of systems, but of reality itself. The basic procedure employed by the General Systems researcher is inductive, involving an examination of various systems and the formulation of statements about the regularities that have been observed to hold. The goal of GST is to organize parsimoniously the knowledge of a diversified domain of events by the use of a few general principles.

General Systems Theory seeks the establishment of a second order theory. The integration of the generally acknowledged findings in the many specialized first order disciplines into a unified optimally consistent and useful framework in which their particular propositions become mutually reinforcing as descriptions and explorations of reality with a rationally knowable, overarching species of order, is one stated aim of this mode of thought (178, p. 55). Bertalanffy (28, p. 48) observes that a unitary conception of the world should be based not upon the possibly futile and certainly improbable hope of finally reducing all levels of reality to a common ground, such as the level of physics, but rather on the isomorphy of laws in different fields. Reality evidences structural uniformities which manifest themselves by isomorphic traces of order in different levels of realms. Models, principles, and laws exist which apply to generalized systems irrespective of their particular kind of element, or the forces involved.

1. Criticisms of GST

General Systems Theory has been criticized on a number of grounds. One of these is that its primarily inductive approach lacks explanatory value. Bertalanffy (28, p. 36) acknowledges this criticism but asks that we consider that there are degrees of scientific explanation; in complex and theoretically less well-developed fields we may have to be satisfied with 'explanation in principle.' The suggestion is made by Bertalanffy that explanation in principle is better than none at all, and that when we are able to insert the necessary parameters, system-theoretical explanation in principle becomes a theory, similar in structure to those of physics.

Another type of criticism regards General Systems Theory as an 'irrelevant distraction' involving nothing but pure common sense. Several geographers have expressed this view of GST. Chisholm (65, p. 48) states, for example, that it is 'premature and unnecessary to dignify the search for generality' with a formal title. Langton (177) feels that the general and general theoretic components of General Systems Theory have received disproportionate attention, and the concept of system, including all its formally defined properties, not enough. Similarly, Smalley and Vita-Finzi (276) see GST as unnecessary in the earth sciences, lending confusion rather than clarification to empirical investigations.

2. Isomorphism

There are two additional objections to General Systems Theory, both of which involve the use of isomorphism. The first objection is that much of the generality observed is based upon the trivial fact that mathematics of some sort can be applied to many different sorts of problems. The law of exponential growth, for example, is applicable to a wide range of varied phenomena. This broad applicability, the criticism runs, has no more significance than the fact that elementary arithmetic may be applied to all objects, that two plus two make four. The second objection warns that General Systems Theory may end up in meaningless analogies. Buck (50) has dealt at length with what he has called the emptiness of analogies. His attitude is that, suppose we do find an analogy or formal identity in two systems, what does this imply? Bertalanffy (28, pp. 35-6) has responded to this criticism by stating that GST is not a search for vague and superficial analogies. Analogies, as such, are of little value since, besides similarities between phenomena, dissimilarities can always be found as well. Isomorphism is more than mere analogy. It is a consequence of the fact that, in certain respects, corresponding abstractions and conceptual models can be applied to different phenomena. Isomorphism stresses the similarity of certain theoretical relationships within systems, not of the absolute properties of elements of systems.

Fein (107) observes that we often substitute phrases from one discipline into another and make analogies in one discipline drawn from another. Rather than being a philosophically naive activity, this may be an implicit recognition of the commonness of our structures of knowledge, and hence their universality. Thermodynamics, for example, has become an attractive model for the study of demographic problems. This is an illustration of the adoption of a developed mathematical structure to aspects of both physical and social behavior. 'The conclusion is not that people act like molecules. This simplistic interpretation misses the essence of the argument. . . . Rather it may be asserted that a similar organization of knowledge can apply to two different classes of aggregate behavior' (107, p. 1379). In certain well-defined circumstances, both people and molecules may be perceived to behave as aggregates and it is possible to use common functional relationships. Although these relationships involve abstractions dealing with the physical behavior of a gas, they are not limited phenomenologically. 'On the contrary, the fundamental postulate here is that structures, saying whatever they do about our perceptions, may be quite general' (107, p. 1379).

Rapoport writes (245, p. 30) that the insights derived from speculations by perceived analogies function somewhat like education; they reveal to the intelligent and conceal from the stupid the extent of their own ignorance. The existence of laws of similar structure in different fields makes possible the use of models which are simpler or better known, for more complicated and less manageable phenomena. Isomorphisms that may be observed between phenomena serve as working hypotheses. If it happens that several phenomena of widely different content can be described by the same mathematical formulation, both simplification and generalization have been effected, and perhaps some hint as to the processes underlying the organizational similarity will have been revealed. The use of isomorphism is possible in science because the same dynamical or functional properties can be seen to be exhibited by large classes of systems of the utmost diversity. Two systems which are physically different but dynamically equivalent are called analogues of one another.

Bertalanffy (28, pp. 80-8) lists three prerequisites for the existence of isomorphisms in different fields and sciences. He further observes that isomorphism apparently rests in our cognition on the one hand and in reality on the other. First, the number of simple mathematical expressions which will be preferably applied to describe natural phenomena is limited. For this reason, laws identical in structure will appear in intrinsically different fields. Second, science is possible. The presupposition for this is that order exists in reality itself. The existence of science proves that it is possible to express certain traits of order in reality by conceptual constructs. All scientific laws merely represent abstractions and idealizations expressing certain aspects of reality. Third, the parallelism of general conceptions or even special laws in different fields therefore is a consequence of the fact that these are concerned with systems, and certain general principles apply to systems irrespective of their nature.

Bertalanffy (28, p. 85) goes on to distinguish three levels in the knowledge of phenomena. The first level involves analogies, superficial similarities of phenomena which correspond neither in their causal factors nor in their relevant laws. At the second level are homologies. These are present when the efficient factors are different, but the respective laws are formally identical. Homologies are of considerable importance as conceptual models in science and are the concern of a general systems theory. A system must exhibit certain general characteristics of systems, irrespective of what elements it is composed of. Logical homology not only makes possible isomorphy in science, but as a conceptual model has the capability of leading to the eventual explanation of phenomena. The third level is explanation, the statement of specific conditions and laws that are valid for a class of objects. Analogies are scientifically worthless. Homologies, in contrast, often present valuable models; they are formal correspondences founded in reality. Explanation is the highest form of knowledge and may perhaps be attained through a rigorous use of homology. It is crucial, then, to make use of the formal identities that may be observed between phenomena, while excluding analogies that merely obfuscate our understanding. The recognition of the structural uniformities of different levels of reality is a large step toward the recognition of the unity of knowledge.

The isomorphism between two or more phenomena is manifest in the existence of a common mathematical structure through which they can be related. Thom (306, p. 72) comments that this agreement between phenomena and a mathematical structure, observed in numerous disciplines, raises a classical problem of epistemology. He notes that three types of answers may be suggested. The first attributes the agreement to a pre-established harmony between mathematics and reality. This is the Pythagorean viewpoint: 'God ever geometrizes.' The second is based on the principle that the phenomena are governed by a condition of local equilibrium, or more precisely are solutions of an externality problem. The third, which Thom advocates, explains the origin of mathematical structure and its morphological repetitions by a genericity assumption. In every circumstance, nature realizes the local morphology which is the least complex possible with respect to the given local initial data. The first answer is pure metaphysics. The second is the only one which may be considered as strictly scientific, as it can sometimes be verified on quantitative models. The third answer is intermediate between science and metaphysics and takes a much more flexible viewpoint than the second, while not being incompatible with it.

There is no simple answer to the question of whether the existence of observed isomorphisms is a function of commonalities in the organization of a broad range of phenomena or is, rather, an artifact of elementary mathematical consequences. As noted above, isomorphisms function as effective working hypotheses that are useful in that they not only facilitate an interdisciplinary cross-fertilization of ideas, but they also provide empirical support to those who seek to develop the philosophical position of the underlying unity of all knowledge. This being so, it is argued that the use of isomorphisms should not be dismissed on the basis of being metaphysical or the results of a triviality in the mathematical formulation employed. Used with discrimination, isomorphisms can be employed as useful tools in formulating new theories, modelling the real world, and advancing the current body of knowledge.

D. THE STUDY OF SYSTEMS: CONCLUSION

The typology of systems approaches that has been presented must be considered as but one of a number of alternative schemata. Although in no way a definitive or absolute classification, the distinctions outlined provide a common set of terms that may be utilized in the subsequent discussion of systems approaches in geography. It is important to recognize, further, that the categories employed are not mutually exclusive. There is considerable overlap between the three approaches outlined, particularly between systems theory and GST. The latter two are complementary and may be regarded as components of a general systems (as distinct from General Systems) approach, the concern of which is both general systems and general theories.

In spite of both the somewhat extravagant claims of its proponents and its minimal number of tangible accomplishments thus far, General Systems Theory cannot with justification be dismissed as a viable component of the methodology of science. If one accepts some conventional definition of science, such as the activity of acquiring knowledge concerning general truths or the operation of general laws, GST, with its goal of the development of a unified theoretical framework for knowledge through the identification of broad commonalities in phenomena and processes, would seem to be a potentially useful methodology. Moreover, the epistemological and philosophical bases to this approach stress an integrative view of the world and of mankind's organized knowledge of that world. Such a view may help to provide a general context or perspective for the individual contributions of the various analytical branches of learning. A general context, in turn, is not superfluous to an individual discipline, but a useful vehicle by which at strategic points in its development it may assess its direction and progress, both internally and in relation to other disciplines.

The preceding discussion of systems approaches has raised a number of points relating to the definition and interpretation of science. Further, both the general systems approach and geography have been introduced within the present chapter in a manner that makes the assumption that it is possible in some way to relate each to science. For these reasons, the purpose of this discussion may now best be served by pausing to examine some of these issues.

IV. Science

In general, two distinct perceptions of the role and function of science may be identified: an idealized or normative view, and another that more closely approximates the reality of the situation. The former view regards science as a nomothetic endeavor; its goal is the formulation of generalized explanations of broad classes of phenomena, as distinct from the observation and description of individual events. The latter view corresponds to Kuhn's (174, chapter 3) concept of 'normal science' and involves the steady extension of the scope and precision of scientific knowledge through the comparison of facts with the predictions derived from the accepted body of theory.

The distinction between these perceptions of science has two sources, the first based upon a set of real distinguishing characteristics and the second based upon a spurious dichotomy. The first distinction involves the notion of the level of maturity of science. In the course of its development, science may be observed to proceed through three levels: a descriptive level in which relevant data are recognized and accumulated, a classificatory level in which a framework for the organization of data is established, and finally, an explanatory–predictive level in which a thorough knowledge of the processes underlying observable phenomena is sought. The latter subsumes the approaches of the former two, and is the idealized form of science. With maturity, not in terms of age but, rather, of epistemological sophistication, comes the capability of attaining this highest level. The three levels are not, of course, necessarily mutually exclusive. A given field is likely to engage simultaneously in all three.

The second, spurious, distinction between the two views of science outlined above is based upon the strict dichotomization of the analytic and synthetic modes of thought. That is to say, the search for general laws is sometimes incorrectly attributed solely to the synthetic mode of thought, while the extension of factual knowledge is sometimes perceived as the only product of the analytic mode. It can be strongly argued that there is, in fact, no logical basis for this polarization, which has resulted from the development of the most extreme aspects of each position. Much of this intellectual extremism has been the direct consequence of the persuasive attempts of such General Systems theorists as Bertalanffy (28), Boulding (40), and Rapoport (245), who have improperly equated analysis with reductionism and empiricism, and synthesis with holism and theory construction.

A. ANALYSIS AND SYNTHESIS

Before exploring the relationship of the analytic mode to the synthetic mode, it may be instructive to consider each separately.

1. The Analytic Mode

The analytic mode is perhaps the most widely employed scientific methodology. The utility of the analytic mode may be attributed to two factors. First, it can perform the important task of eliminating unverified and unverifiable speculation (178, p. 3). This is of particular necessity in the case of a non-mature science that has not developed a viable paradigm. Second, analysis enables a researcher to investigate manageable units of reality as opposed to the seemingly undifferentiated complexity of the world. In order to study some hypothesized causal relation free of the disturbance of other facts, all those suspected of having some influence are held constant. The basic assumption of this method is, thus, that all disturbing phenomena can be eliminated and the relation of interest studied in isolation (244, p. xiv). This is, of course, a simplifying assumption, but oftentimes a very necessary one. By establishing several pairs of such relations, arrived at in isolation, it is further assumed that they can be combined into a more general causality law. Such two variable linear causal interaction schemes emerged as one of the earliest, and principal, modes of scientific explanation. This mode has with some justification been the object of criticism by General Systems theorists, among others, on the grounds that it ignores the larger interconnections which may be decisive for the understanding of phenomena. Two observations on this criticism may be offered, however. First, a suitable method of handling the larger interconnections is often hard to discover, even among the General Systems concepts. Second, contemporary analytic activity, with its highly developed probabilistic and multivariate tools, is misrepresented to a considerable degree by reference to this simple model of explanation.

2. The Synthetic Mode

The synthetic mode generally does not involve the unwarranted speculation sometimes attributed to it but, rather, a method of approaching both empirical data and theory so as to furnish new avenues toward constructive consideration of substantive issues and phenomena relating to man and his physical environment. While the analytic method has emphasized the development of precise techniques for making empirical observations and for the analysis of data, it is the role of the synthetic mode to attempt to view the data in new ways that will promote the discovery of general relationships among them. We appear to be part of an interconnected system of nature, and unless systematic theories of the patterns of interconnections are developed, a full undestanding of this complexity may elude us.

3. The Interrelationship

Much of the criticism of the analytic mode on the part of General Systems theorists may be ascribed to an ecological fallacy that has been accepted in their view of the sociology of science. While the discovery of broad commonalities and general laws among all phenomena and processes is, one may argue, one of the ultimate aims of mankind's search for knowledge, certain proponents of the synthetic mode fail to distinguish between the role of the individual and that of the group. Virtually every theoretical synthesis that has occurred in science has involved both the steady accumulation of data gathered through precise observation and experimentation, and the insight gained from taking a larger view of the relations suggested by the data. When science, the product of the activity of scientists, advances through a theoretical synthesis, it does so as the result of many individual analytical contributions. Thus, analysis and synthesis are not antithetical concepts but are, rather, complements which together provide science with its most productive methodology.

B. SPECIALIZATION AND GENERALIZATION

A second pair of dichotomous views is sometimes identified in science – specialization and generalization. Since a relation to the analysis–synthesis dichotomy is often ascribed to this pair, and since a theme developed in the succeeding chapters involves the utility of generalization in treating the substance and concepts of geography, a brief consideration of these views may prove beneficial.

Laszlo (178, p. 8) tells us that all coherent and systematic theories of the empirical world are based on two 'primary presuppositions' shared by all empirical scientists: first, the world exists; and second, the world is intelligibly ordered, that is, open to rational inquiry. The acceptance of these very basic postulates permits the scientist to begin the rational mapping of the empirical world and the task of theory construction. Laszlo continues by observing that there are also two secondary presuppositions:

one, that the world is intelligibly ordered in special domains; *or*, two, that the world is intelligibly ordered as a whole. This latter pair of postulates is the boundary that separates the specialist and the generalist. One or the other presupposition is made by every investigator, but those that make the first one often refuse to admit that it is a presupposition ; they tend to look upon the intelligibility of the phenomena that they investigate as a 'fact of nature.' On the other hand, the second presupposition is usually considered to be in need of demonstration. The specialist limits the scope of his inquiry to a carefully isolated class of events and assumes that his domain is intelligible, but often refuses to consider that it may be interwoven, with equal intelligibility, with other domains of existence. The generalist makes this assumption and holds that a knowledge of phenomena and processes in any one particular domain may be fully realized when brought into conjunction with a knowledge of other domains. As Whitehead (353, chapter 1) has observed, 'The point is that every proposition refers to a universe exhibiting some general systematic metaphysical character. . . . Thus, every proposition posing a fact must, in its complete analysis, propose the general character of the universe required for that fact.'

The prominent place in science of positivism has caused the burden of proof to be unjustly distributed among the two secondary presuppositions. The first is considered by most scientists as a 'given'; the second is accused of being metaphysical (which Laszlo suggests connotes untenable or, at best, based on unconfirmed speculation) and in need of empirical verification. Since both presuppositions rest on the primary postulates of the existence and intelligibility of the world, the demands for proof must be distributed equally. Neither special nor general theories have a privileged status as a necessary truth (178, p. 9).

Much of the acceptance of the specialized form of inquiry characteristic of positivism is based upon the relatively higher degree of precision and attention to detail that can be achieved. Yet, this preference is based on the assumption that special order is intelligible in itself, or at least that it is not rendered significantly more intelligible by consideration of general, more encompassing orders. The truth of this assumption is, in turn, dependent on the prior choice between the secondary postulates. If the world is intelligibly ordered as a whole, then the more regions of order that are disclosed by a theory, the more that theory is divested of the personal bias of the investigator (178, p. 10). Thus, both secondary presuppositions may be regarded as of equal heuristic merit; attempts at general theories are intrinsically neither better nor worse than attempts at the formulation of special theories.

Considering the effects of specialization, Boulding (40, p. 24) notes that knowledge is a function of human organisms and social organization; it does not exist and grow in the abstract. The manner in which knowledge grows is through the receipt of meaningful information by a 'knower'. Science is, thus, the knowledge that can be profitably discussed by scientists in their role as scientists. The crisis of science arises today because of the increasing difficulty of such profitable talk among scientists. Specialization has made communication among disciplines and among sub-disciplines increasingly difficult, causing isolated subcultures with only tenuous lines of communication between them (40, p. 4). In the course of specialization, not only the domain of science but the 'receptors' of information, the scientists, become specialized. 'Hence physicists only talk to physicists – still worse, nuclear physicists talk only to nuclear physicists' (40, p. 4). The more that science breaks into sub-groups, and the less the amount of cross-communication that takes place, the more likely that the total growth of knowledge is being inhibited. 'Specialized deafness' is the result.

Specialization is, to a large extent, responsible for the proliferation of disciplines, sub-disciplines, and fields which has threatened to fractionate and compartmentalize the scientific community into mutually isolated enclaves unable to communicate with each other. The unfortunate consequences of these 'specialty barriers' that have arisen is that knowledge, instead of being pursued in depth and integrated in breadth, is pursued in depth and relative isolation (180, p. 4). The specialist concentrates largely on detail and tends to disregard the wider structure which gives it context. The scientist who takes a generalist's view of the complexity of the real world concentrates, rather, on structure and organization at all levels of magnitude and fits detail into its general framework. He attempts to discern relationships and situations, believing that some knowledge of connected complexity is preferable to an even more detailed knowledge of atomized simplicity, if it is connected complexity with which we are surrounded in nature (180, p. 13).

The direct manifestation of specialization can be observed in the organization of the modern university. Originally founded to pursue the study of universal knowledge, the university has laid aside its main focus of the unity of that knowledge and has contributed to its compartmentalization. The concept of unity of knowledge has become subordinate to the achievement of a neat structure for the rapidly proliferating sub-specialties. The institutions known as universities may be seen to be both a cause and an effect of the fragmentation of knowledge, and may more realistically be viewed as multiversities and diversities. Waddington's (329, p. 410) observation clearly expresses this view.

> In its old elitist day the university used to provide a general education for gentlemen. This was supposed to be an introduction to all things that such a person should know something about if he wanted to understand society. We have swept that away entirely and concentrated on training specialists, either because there is no agreement on the nature of civilization, or because we have no time to consider it. But I believe that universities are failing in their responsibilities by not providing some sort of system for making the students acquainted with the developments of modern thought.

C. THEORETICAL SCIENCE

As noted above, all science begins as description; relevant facts are chosen and then organized. Advances come when individual discoveries are fitted together into theoretical structures and tested. It is only within the last decade or two that geography has made appreciable progress toward becoming a theoretical science. With optimism that the discipline will continue its development, this brief examination of science is concluded by considering the nature of theoretical science.

Science seeks not merely an account of the world but a simplified, systematized account. The conceptual simplicity that is sought is measured not only by the number of axioms employed but also by the simplicity of the concepts utilized. The goal of science, then, involves both explanation and the efficient organization of the propositions of which science is composed; an economical predicate basis or set of concepts is sought. Theory is the means through which this economy and simplicity may be achieved. Theory, the *raison d'être* of which is economy of thought, makes it possible to rest an immense superstructure of knowledge on a minute foundation of propositions. This is in contrast to disciplines, such as chemistry, which do possess a limited theoretical code of orderliness, and where a theoretical edifice would consist, rather, of an enormous foundation supporting a small superstructure (113, p. 15). In those phenomenological domains in which novelty is an immanent feature, a theoretical edifice, even if feasible, would be uneconomical. Georgescu-Roegen (113, pp. 114-16) uses the concept of 'novelty by combination' to account for the separation of chemistry from physics. Chemistry is interested in how the finite number of chemical elements combine to produce various qualities of substances, most of which cannot be deduced from the simple properties of the elements involved. Almost every new chemical compound that is synthesized is a novelty in some respect, from the viewpoint of extant knowledge.[11] Thus, no theoretical organization is really possible. Further, many fields that are struggling with patchy knowledge that has not been unified into a single theoretical body deal with variables that are quantified qualities which empirical science, with its emphasis upon quantification and measurement, is unable to cope with.

Georgescu-Roegen (113, pp. 116-19) continues this trend of thought in the definition of rational phenomena of the first, second, and third orders. Rational phenomena of the first order are those that can be logically deduced from other propositions; those of the second order are phenomena that cannot be known unless actually observed; those of the third order represent phenomena that cannot be predicted even after the same combination has been observed several times. The latter group of phenomena is unpredictable primarily as a function of the added dimensions of novelty that appear as we move from the inert to the organic and to the supraorganic domains. It is rational phenomena of the third order with which social science is primarily concerned and to which its methodology must be adapted.

Theoretical science, then, is a catalog which lists known propositions (as distinct from facts) in a logical (as distinct from taxonomic) order; it is the logically ordered, economical organization of knowledge. Although the volume of factual information about the real world has expanded continuously, its number has mattered less because of the growing number of domains that have been brought under the power of theoretical understanding (113, pp. 26-7). Medawar writes that 'in all sciences we are progressively relieved of the burden of singular instances, the tyranny of the particular. We need no longer record the fall of every apple.'[12]

Georgescu-Roegen (113, p. 37) makes the analogy of theoretical science as a living organism, a purposive mechanism that reproduces, grows, and preserves itself. Anatomically, theoretical science is logically and economically ordered knowledge; a catalog of facts is not science. Physiologically, it represents a continuous secretion of experimental suggestions which are tested and organically integrated into the anatomy of science. Theoretical science is not only a more convenient storage place of knowledge but also a more efficient instrument than crude empiricism in expanding knowledge. New propositions can be derived from the logical foundation of a science without interruption. One function of theoretical science is precisely the continuous derivation of new propositions; new facts are continuously created from old. And teleologically, it may be likened to an organism in search of new knowledge.

Theoretical science has come to recognize its own limitations, particularly to recognize that the goal of perfect explanation can rarely be achieved. The view that science must be free of any contradiction is no longer a viable position. Physics, the most well-developed model of science, taught us that we must not insist on molding actuality into a noncontradictory framework (113, p. 9). This has been formalized by Niels Bohr (38, p. 40) into an epistemological tenet, the Principle of Complementarity: 'Only the totality of the phenomena exhausts the possible information about the objects.' Two theories may be mutually contradictory, but each is noncontradictory within itself and thus both must be accepted side by side. The corpuscular and wave theories of the electron provided the impetus for Bohr's formulation.[13,14] Bridgman (44, p. 118) summarizes this view: 'the only sort of theory possible is a partial theory of limited aspects of the whole.' That is to say, in most instances no single theory suffices to explain an event.

By extension, the Principle of Complementarity may be applied to the analytic–synthetic dichotomy discussed above. Science must learn to recognize that neither of the two epistemological positions alone is sufficient as a basis for science. The endeavors of seeking simplification (in the sense of reduction to component parts) and of acknowledging complexity are contradictory, in a sense, but also complementary. The basis of knowledge cannot be ascribed to either the whole alone or to the parts by themselves.

Notes

[1] The internal and external relationships of an area of knowledge may together be regarded as a framework (137) or a conceptual structure (96). A framework is a high-order proposition, as distinct from a simple low-order description, concerning the real world and indicates both what part of reality should be examined and which elements in the broad expanse of complexity ought to command prior attention. The specific nature of the framework therefore influences the low-order propositions that are made about the real world. There are several important aspects of frameworks or conceptual structures that need to be considered. First, frameworks cannot be proved or validated but represent only working assumptions (216) that improve the internal coherence and implicative structure in our understanding of the real world (63). Second, utilization of a framework necessitates value judgments of one sort or another. Since frameworks are based largely on concepts which are, in turn, neither true nor false but only more or less useful (97), selectivity is inevitable and so, too, is subjectivity. The idea of a value-free science must be recognized as a myth (281). Third, frameworks are evolutionary in nature, being continuously refined or discarded in whole or in part as new knowledge suggests more useful or appropriate ways of investigating and organizing reality. Fourth, since the real world is highly complex, there is some advantage in a broad framework that explicitly relates the developments in one area of knowledge to those occurring in other areas. Each field contributes to the advancement of science. At the same time, however, the course of science as a whole determines the progress of its parts (4).

[2] The role of the deductive mode in bringing about the majority of significant contributions to human knowledge was explained to me by the philosopher of science, C.A. Hooker.

[3] This notion is expanded on in chapter 4.

[4] Only the latter, the specification of both the sequence of events and the mechanism, qualifies as a rigorous usage of the term process. The two former usages are both incomplete and, thus, informal.

[5] This is yet another manifestation of the scale problem in geography. What appears as process at one 'temporal scale' appears as structure at another.

[6] While similar in most respects, the views of Harvey and Schaefer are dissimilar in that the former strove for unification and synthesis of these concepts while the latter did not.

[7] An interpretation of the origins of the spatial tradition in geography is undertaken by Pooler (241).

[8] There is no intent to suggest that these have been the sole candidates for organizing frameworks within the discipline. For example, the concepts of environment, ecology, and ecosystem, which are implicit in much of the regional framework (18; 263; 326), have recently reappeared, independently of any explicit notions of areal variation, as candidates (153; 295; 322).

[9] See Hartshorne (144; 145) and Schaefer (264). This debate has been continued in the recent geographical literature by Bunge (55; 53), Sack (259; 260; 261; 262), Guelke (134), Warntz (333; 337; 343; 347), Dolphin (92), Harvey (147) and others.

[10] These concepts will be illustrated subsequently. As they are now fairly common terms, no attempt will be made to define them at this time. Virtually any discussion of 'systems analysis', 'systems theory', or 'General Systems Theory'

may be referred to by the reader if clarification is required. See, for example (28; 147; 68; 208).

[11] The view of chemistry advanced by Georgescu-Roegen employs a limited definition of the simple properties of elements, excluding, for example, the configuration and number of electrons of an element.

[12] P.B. Medawar, *The Listener*, May 18, 1967, p. 647; quoted in Georgescu-Roegen (113, p. 27).

[13] These two theories of the electron may be viewed as dialectical. Dialectical concepts are those that violate the Principle of Contradiction: 'B cannot be both A and non-A.' In certain instances, at least, we must accept that B is both A and non-A (113, p. 46).

[14] The contradiction which Bohr sought to resolve was abrogated by the formulation of quantum theory.

2

Geography

Having set the context for this discussion, and having examined several important philosophical and epistemological issues in science, we turn now to reflect upon the discipline of geography. The substance of the discipline and its phenomenal and methodological boundaries are examined first. Next, contemporary geography is appraised in view of the observations made in chapter 1 on the nature of science. Finally, the role of theory in science and, specifically, within geography is considered.

The present chapter is an exploration of the logical and philosophical validity not only of the discipline of geography but also of the practice of geography at a theoretical level, for it is at that level that the systems framework, in its several forms, has been introduced into the discipline.

I. On the Nature of Geography

Before proceeding to an examination of systems approaches in geography, it is first necessary to explore the nature of the discipline and to delimit its realm of concern. What is the nature of the subject matter that geography studies? Where does geographic research begin, and where does it end? Are there purely geographic phenomena? Does geography have a distinguishable unity and, if so, what accounts for or defines that unity? These are crucial questions; it is the answers to them that determine whether it is indeed possible for a theoretical, law-seeking discipline of geography to exist.

In attempting to delimit the scope of geographic inquiry there are several fundamental questions that need to be considered. First, should the *geo* in geography be strictly interpreted? In other words, is geography to be regarded as a science of *terrestrial* space or as a science of space, in its generic sense? There is, of course, a strong tradition underlying the former alternative, manifest in the etymology of the term 'geography', in the ancient partitioning of knowledge between geography and astronomy, and in the conventional definitions of geography which involve some variation of the 'earth as the home of man' theme. A second consideration involves the opposite end of the continuum of scale. Does the study of spatial relations among small-scale phenomena at some point cease to be regarded as geography and become physics or biology? A traditional interpretation of geography would suggest that events and processes below the level at which direct interaction with the human organism may be perceived lie outside of the discipline.

A cogent case may be made, however, for an interpretation of the scope of geography that transcends these conventional views of the discipline. The range of human awareness is a direct function of available technology and is, thus, characterized by highly fluid boundaries. As a result of recent technological advances man's perceptual capabilities have undergone a marked expansion; he has become both an observer of, and a direct participant in, events that occur at extra-terrestrial and at micro-terrestrial scales. As the boundaries of geographic enquiry that have been employed *de facto* in the contemporary period have their origins in the knowledge and technology of classical antiquity, there may be some merit in taking a broader view of the discipline in light of the accumulated advances in human awareness and understanding that have marked the past 2500 years.

It is this expanded view of the discipline of geography that is taken in the present discussion. This position may be justified on several bases. First, geography may still be properly regarded as the study of 'the home of man' or of the relationships between 'man and his environment'; environment is broadly construed to include the entire range of human awareness, from the astronomic to the sub-molecular. In an age when man is walking on the moon and accelerating atomic particles, such an interpretation of environment does not seem inappropriate. Second, recent experience has demonstrated the utility of an inter- or trans-disciplinary approach to the acquisition of knowledge (4; 218; 332). In addition to providing its own original insights, through the theme of terrestrial space, geography has served to integrate the knowledge of man and his meso-scale environment that has been acquired in such fields as anthropology, economics, geology, and plant science. There is no reason to believe that the discipline cannot function in a similar manner by applying its perspective of generic spatial relations to such fields as astronomy, physics, biology, and chemistry.

This view of the discipline reflects a 'geography is what geographers do' interpretation which implies that geography is not defined so much by the subject matter that it examines, in Wilson's (360) terminology its system-of-interest, as by its methodology – the concepts, tools, and viewpoint employed. This interpretation is in no way a radical departure from the more conventional approaches to the discipline. Geographers will continue to investigate those problems that have traditionally concerned them. The point is that it is a valid endeavor for them to expand their horizons so as to include the study of spatial relations at all scales within the scope of human awareness. Thus, such topics as the topology of the Martian canal network (350) or the location of orbiting man-made space debris (56, pp. 406-411) are valid geographical problems. Examples of geographic inquiry at the lower end of the scale continuum include the investigations of Woldenberg (365; 370) into the spatial form of human and animal organs and of Eichenbaum and Gale (100) into hexagonal tessellation of algae. Contributions to the spatial analysis of small scale phenomena are also to be found in formal disciplines other than geography, supporting the above contention that geography is an integrating science. The series of symposia edited by C.H. Waddington, *Towards a Theoretical Biology* (330) contains many examples of geographic problems of morphology and location in biological phenomena.[1] D'Arcy Thompson's *On Growth and Form* (309) has had a profound impact on the spatial thinking of many geographers, and Stevens (288) presents many instances of spatial commonalities at all scales of nature. He indicates, further, that many phenomena at one scale are spatial analogues of those at a different scale, and operate according to similar types of spatial process.

Geography can be, and must be, defined in the broadest possible manner: it is the science of space-occupying processes and events.[2] The concerns of this science encompass not only the set of all places on the earth's surface, which necessarily have spatial properties (absolute or relative location) and are capable of having non-spatial properties (climatic, economic, etc.), but also those phenomena and processes which are conceptualized primarily in terms of non-spatial characteristics and, in addition, are in some manner distributed in space. It follows from this latter point that geography shares a great deal of its subject matter with other disciplines, yet it approaches the subject matter from a viewpoint that the other disciplines do not – the spatial view. Thus, because every process or event has some sort of spatial manifestation, geography may be conceptualized as a discipline which transcends the conceptual and empirical boundaries of compartmentalized knowledge about reality, and integrates that knowledge in a useful and meaningful way. Geography is, indeed, an integrating science (52, p. 329) and the theme around which the knowledge of the processes and objects of the real world is integrated is that of space.

The substance of geography *may* include all processes and objects that occur within the range of human perception with one class of exceptions. This consists of those phenomena and processes that do not have explicit spatial properties or, more often, that have spatial properties which are held by convention to be of minor significance. The exceptionalism of this class is, in general, hard to justify completely for the boundaries of geography are not only wide; they are permeable. For example, psychological processes such as learning, cognition, and perception have been conventionally investigated without reference to explicit spatial components and thus have been generally considered beyond the domain of geography. Recently, however, the boundary has begun to blur as both geographers and psychologists have sought to explore the indirect links between the two fields. Attempts have been made to assess the impact of the spatial configurations of the environment upon perception, attitude formation, and learning and, conversely, the role of perception and other mental processes upon spatial behavior has been investigated.[3] Bjorklund and Philbrick (32), for example, have delved into the direct links between mental process and the spatial attributes of human behavior. The point is that while links may exist between psychological processes and space or spatial behavior, spatial properties are not the primary focus of psychology and, thus, psychological processes and related phenomena have been regarded as beyond the substance of geography. However, as indicated, a spatial viewpoint and a psychological viewpoint are not totally distinct and in combination may enhance the sum of knowledge about the real world.

The discipline of economics, too, presents many examples of phenomena that are investigated aspatially: price theory, economic growth, welfare economics, the theory of the firm, to cite a few. Again, the spatial properties common to some economic sub-specialties have been developed, resulting in what has come to be known as economic geography and regional science. Similarly, the study of English literature usually involves no spatial analysis, yet a geography of English literature brings a spatial view to an aspatial discipline. We may observe, then, that the subject matter of geography can be extended into the realm of 'non-spatial' disciplines that, by virtue of the particular viewpoint adopted, ignore the medium of space. All 'non-spatial' phenomena have a direct or indirect spatial manifestation, an absolute or relative location of either the phenomena themselves or of their effects, that exerts an influence on the other objects or processes involved; the effects of space can rarely be held constant. This was implicitly recognized in the early European universities where geography was the basis of both physical and social science, prior to the emergence of specialization and its fragmentation of the existing unity of knowledge. Geographers are, then, justified in examining any phenomenon or process which they feel can be elucidated by the application of a spatial perspective.

It should be quite evident from the above discussion that there are many more individuals actively engaging in geographic research than there are formally trained geographers. One could argue that those persons in 'non-spatial' disciplines that are engaged in investigations of the spatial aspects of phenomena are performing geographic inquiry. Conversely, geographers have had an impact on other disciplines through the consideration of the spatial properties of 'non-spatial' phenomena. These facts lend strong support to the position that geography is an integrating science, a common ground that is able to unite the efforts of many scholars. Further, these facts suggest that there is indeed a basis for a general spatial science that transcends the artificial boundaries of specialized disciplines.

The broad definition of the subject matter of geography advocated above has anticipated the answer to the question as to whether uniquely geographic phenomena exist. Any phenomena or processes that may be considered as possessing spatial properties and that are within the scale of man's observation are substance for geography. It follows that because geography transcends artificial boundaries imposed upon knowledge, there can be no uniquely geographic phenomena in the sense that unique phenomena can exist when only non-spatial properties are considered.

Thus, it is possible to conceive of phenomena that may be classified as uniquely geological, economic, or psychological, but not of phenomena that are uniquely geographic. It must be understood that the capability of classifying a phenomenon as purely geological or economic is contingent upon ignoring the underlying spatial components that may exist; 'uniquely' geological or economic phenomena may also be 'non-uniquely' geographic.[4]

A related problem is that of whether there exists uniquely geographic explanation. That is to say, are there objects or processes whose spatial or non-spatial properties may be explained on the basis of space or spatial relations alone? This is an extremely difficult problem and one that has been debated in the geographic literature for many years. Schaefer (264) holds the view that geography is essentially morphological, and that geographic explanation can be achieved through the application of the laws of geometry. Sack (261) takes the view that no explanation can be called peculiarly geographic, and that with trivial exceptions all explanations except those of synthetic geometry can be used to answer aspects of geographic questions.

The view adopted in the present work is that a degree of purely spatial explanation does exist. Space, in its Euclidean metrical sense, appears to exert an influence upon both the distribution of phenomena and upon their non-spatial properties; there are numerous location and interaction studies that have furnished empirical evidence for this view.[5,6] Many non-spatial principles, and many phenomena that do not have explicit spatial properties can also be demonstrated to contribute to geographic explanation. It is argued, however, that while non-spatial factors can contribute to geographic explanation, they are not able to make their contribution independently of spatial properties. Spatial properties, then, are able to provide a first, and often very close, approximation of explanation, independently of non-spatial properties. Non-spatial properties are able to enhance the explanation of spatial relationships only where spatial properties are already being employed in explanation. Sack's view that all forms of explanation can be used to answer aspects of geographic questions is an implicit recognition of the complexity of phenomena in the real world; it is often difficult to separate the spatial component from 'non-spatial' processes. For example, does not commodity price, which is determined by 'aspatial' supply and demand functions, vary over space, and according to timing mechanisms which are largely introduced by locational considerations?[7] There is a circular and cumulative feedback loop operative between the spatial and non-spatial aspects of explanation. That is to say, the spatial properties, such as location, of any phenomena can exist independently of non-spatial properties, but are certainly influenced by non-spatial properties. The nature or intensity of the non-spatial properties are, in turn, influenced by absolute or relative location. This implies that it is possible to consider the distribution of phenomena in neutral terms, holding constant their non-spatial properties, and still be able to achieve a degree of explanation. To the extent that this abstraction, the consideration of non-spatial properties neutrally, allows generalizations to be made and widely applicable principles to be introduced, the neglect of non-spatial properties is justified. Perfect explanation of particular phenomena, which is rarely achieved in any event due to the complexity of the world, is thus replaced by a first approximation of explanation that is applicable to broad classes of phenomena. This would seem to meet the criterion of economy proposed by Georgescu-Roegen (113) for a theoretical science.

The type of epistemological position advanced here is open to the criticism of those who would argue that by establishing a broad range for the substance of geography the discipline is reduced to being virtually meaningless. Indeed, one of the arguments proposed by those seeking to remove geography from the university curriculum is based upon the facility with which its subject matter may be allocated to other disciplines. This argument interprets – incorrectly, it is suggested – the transcendent nature of geography as a weakness rather than as a strength.

We now turn specifically to some of these objections to the view of geography advocated here. By way of conclusion to the above discussion, it can be observed that the view of the nature of geography that has been presented is a logical consequence of a philosophical–ontological position that stresses the unity of nature and the unity of knowledge about nature; geography has a distinguishable unity in the sense of a general applicability to all phenomena that manifest themselves spatially.

II. Criticisms of the Spatial View

The above comments present a view of geography that may be characterized by the term 'spatial'. The view developed here transcends the traditional spatial approach, represented by the work of Bunge (55) and Schaefer (264), in the breadth of its scope and its recognition of the complexity inherent in attempting to distinguish between spatial and non-spatial explanation. Yet, the approach of the present work is one that is unmistakably spatial; geography is the science of space, and space may be observed to exert a definite influence on phenomena in terms of both the location of objects and processes, and their non-spatial properties. Sack (259) cites the spatial view as an inversion of the chorological concept, the doctrine advocated by Hartshorne (144; 145), among many others, that a knowledge of the earth can be achieved through the method of areal differentiation. The spatial view, then, holds that the specific relations between phenomena are more important than the nature of the phenomena themselves.

The major source of recent criticism of the spatial view has come from Sack (259; 260; 261; 262); Hartshorne (144), too, has attacked this approach but as his criticisms are both well-known and at times directed *ad hominem* we may safely ignore them here.[8] The primary focus of Sack's criticism is what he terms the spatial separatist theme in geography. Briefly, the spatial separatist theme involves the view that it is possible to identify, separate, and evaluate the effects of space (or distance), either as an independent phenomenon, or as a property of the events examined through spatial analysis. Bunge (55), Schaefer (264), and Warntz (336; 342) are identified as the principal proponents of this approach. Three major instances of the spatial separatist theme are identified by Sack (262). The first involves the claim that there are spatial, non-spatial, and degrees of spatial fact, or that space and the properties of space exist absolutely; Warntz and Schaefer are cited as examples. The second instance revolves around the claim that distinctions can be made about the ability of laws to explain and to predict physical geometric properties of events according to the degree to which their empirical concepts include spatially significant terms; Bunge and Schaefer and their reliance upon geometry and morphology are said to exemplify this approach. The third instance is what Sack terms whole part analysis. This refers to assertions that the significance of spatial properties of an event is determined by the physical geometric properties and arrangements of its parts and their effect upon one another and upon the whole. The general systems approach to geography, in general, and Langton (177) and Eichenbaum and Gale (100), in particular, are cited.

Sack's objections to the view that he terms spatially separatist are based primarily upon the relational concept of space (260). This concept states that every instance of geometric or spatial terms must be connected or related to one or more instance of non-geometric terms. In order to be a useful concept in science, Sack claims, physical space must have named substance referents. As examples of the use of a non-relational concept of space in geography the distance decay hypothesis, central place theory, and von Thunen's law are cited; absolute space, space that is not related to some non-geometric term, is employed in each. Sack presents an argument that is internally consistent in its logic but deficient in its empirical and theoretical validity (242). He is criticizing, solely on the basis of a philosophical position (one which, as we will see below, has been convincingly refuted), formulations that, while admittedly not rigorously developed laws and so open to several forms of criticism, nevertheless have been proven to be of widely applicable theoretical and empirical utility. Additional examples of empirical regularities that are the results of law-seeking deductive–nomological endeavor which have done much to advance our knowledge of the real world but which violate Sack's criterion of the use of named substance referents may be cited.[9] The position held by Sack appears to be untenable in view of the evidence that suggests that a non-relational concept of space is, in most instances, both a necessary and sufficient component of attempts to answer spatial questions.

Sack carries this argument to an extreme position with his claim that a geometric fact may be explained by a non-geometric law (261, pp. 68-9). He cites the problem of the calculation of the hypotenuse of a rectangular living room. What must be sought, in keeping with the relational concept of space, Sack claims, is a law about living rooms rather than about rectangles. Further, since the subject under consideration is the wall of a living room, not the side of a rectangle, one does not enter into the domain of geometry when calculating the diagonal length of the living room. This would seem to be a perversion of the worst sort of the goals of science. Sack is taking extra pains to deal with the particular rather than with the general. If one were to pursue his line of reasoning, it follows that it is necessary to formulate *n* laws to deal with rectangles, where *n* is the number of possible classes of things that can assume a rectangular form. Sack begins with a general law which he does not choose to apply because of the relational concept of space, and thus proceeds from the general to the particular. This is consistent with neither the epistemology nor the methodology of science.

In a broader context, Sack's arguments may be seen to be closely associated with, and to have roots in, the operationist school of thought, a product of the methodological work of the physicist, P.W. Bridgman (43). The central idea of operationism is that the meaning of every scientific term must be specifiable by indicating a definite testing operation that provides a criterion for its application (152, p. 88). Such criteria are often referred to as operational definitions. The implication of this is that a term has meaning only within the range of those empirical situations in which the operational procedure 'defining' it can be performed. As Hempel (152, p. 91) points out, however, a too restrictive operationist construal of the empirical character of science has tended to obscure the systematic and theoretical aspects of scientific concepts and the strong interdependence of concept formation and theory formation. The operationist view obliges us to countenance a proliferation of concepts, distance being but one instance. This not only produces a set of terms that is both practically unmanageable and theoretically endless, but also defeats one of the principal purposes of science, namely the attainment of a simple, systematically unified account of empirical phenomena. Moreover, simplicity in the sense of economy of concepts is an important feature of science; and broadly speaking, the systematic import of the concepts in a theoretically economic system may be said to be stronger than that of the concepts in a less economic theory for the same subject matter. 'Thus, considerations of systematic import militate strongly against the proliferation of concepts called for by the maxim that different operational criteria determine different concepts' (152, p. 94).

Sack continues his argument by claiming that the apparent generality and deductive rigor of many of the spatial formulations in geography have been due to the very omission of the referents of the spatial concepts and to the assertion that physical spatial properties are *a priori* useful. It would appear that Sack has chosen to ignore that the role of science includes the function of economization, of reducing many particular facts to a more manageable number of more widely applicable ones. One way in which this generalization can be accomplished is to view objects and processes in abstraction, to ignore unique characteristics and to focus upon the commonalities of phenomena. The requirement that named substance referents must be employed does not seem to be in accord with the goal of science. Geography is the science of space, and includes road-defined space, river-defined space, and so forth. Just as Bunge (54) argues that locations are not unique, Pooler (242) concludes that the same must be said of distances. As to the assertion that the concept of physical space is *a priori* useful, this assumption would seem to be a reasonable point of departure for geographers at this stage in geographic thought, given the number of instances in which the utility of a non-relational concept of space has been demonstrated.[10] Concluding his discussion of the relational concept of space, Sack notes that it raises objections to the notion that geography is a spatial science apart from other sciences. This is aimed primarily at Schaefer's (264) references to the existence of purely spatial explanation. Geography is a spatial science, even in the sense, discussed above, of providing a degree of explanation based on spatial concepts alone. Geography, however, is not separate from other sciences; it encompasses them and transcends their artificial boundaries, it integrates them from the viewpoint of space.

Having invoked the relational concept of space, Sack (261) goes on to examine specifically the notion that no explanation can be called peculiarly geographic, that with trivial exceptions all explanations except those of synthetic geometry can answer aspects of geographic questions. This question has already been discussed in detail above. It can be emphasized again, however, that many of Sack's objections are negated when the view of geography as an integrating science, as the spatial component of a unified science, is adopted.

One further and related objection is that the explanation of process cannot be inferred from geometric properties alone.[11] That is to say, the explanation of process cannot be achieved when phenomena are examined in neutral terms, spatial properties included but non-spatial properties not regarded. This follows, Sack argues, because geometric properties such as distance, direction, and shape can only produce static laws, while explanation hinges on process, which is dynamic. Bunge (55, p. 211), however, points out that rather than the static and the dynamic being contradictory concepts, they are dual expressions. For example, it is the spatial processes of air masses that produce the spatial structure of Koppen's hypothetical continent and the optimum movement of agricultural products to the center of von Thunen's Isolated State that provided rings of agriculture, and so forth.

Bunge notes that spatial structure can be defined most sharply by interpreting structure as geometrical, which term includes many geographic concepts such as pattern, distance, morphology, shape, slope, local relief, distribution, among others. Thus, patterns and motions, the static and the dynamic, constitute Schaefer's notion of spatial relations (55, p.

213).[12] Where Bunge holds the more limited notion that spatial structure is the result of spatial process, it may be demonstrated that neither structure nor process is uniquely cause or effect of its dual. That is to say, not only is structure the result of process, but process is to a large extent the result of structure. (See chapter 1.) Warntz and Woldenberg (347) use a 'properties of surfaces' methodology to portray general properties relevant to all continuously differentiable geographic surfaces. (The geographical surfaces referred to include those components which Bunge subsumed under the term geometrical, as noted above.) They point out that on any surface 'natural' movements may be identified, and that these natural movements tend to be along steepest slope lines or gradient paths. Stewart (291), Warntz (336), and Coffey (79) have demonstrated the influence of spatial structure upon spatial process in both inter- and intra-metropolitan systems. Warntz and Woldenberg (347), Dutton (94), and Coffey (79) have, further, noted the relationship between the geometry of a system and the intensity of the flow process that is maintained. Geometric structure and process are not independent phenomena, and must be regarded as integrated rather than dichotomized.

In summary, the nature of geography has been considered and the discipline found to be usefully regarded as a spatial science, the substance of which may be broadly defined so as to transcend the artificial boundaries of specialized sub-fields which have compartmentalized knowledge. The objections to the spatial view of geography have been noted to focus on the relational concept of space and the related position that there can be no explanation based solely on the spatial properties of phenomena. However, it has been argued here that a non-relational concept of space is not only a useful one for conducting geographic inquiry, but also one which is in accord with the economy and generality which characterize science. Similarly, the relationship between the static concept of spatial structure and the dynamic concept of spatial process has been identified to be such that a degree of explanation can be achieved from a knowledge of form. In short, there has been nothing found in either the substance or the methodology of the discipline which would preclude the existence of, or the development of, a truly theoretical geography. It is to a consideration specifically of theoretical geography that we now turn.

III. Contemporary Geography

Anuchin (10, p. 43) tells us that periods in which the general absorbs the particular are succeeded by those during which the particular destroys the general and a single science disintegrates into an endless number of branches. Contemporary geography is in such a phase of differentiation. A plethora of sub-fields has arisen where once stood the unified discipline whose domain was the totality of physical and social aspects of existence on the planet earth. From the initial dichotomy of human and physical geography has followed a seemingly endless number of topical specialties, each subdivision based on the non-spatial properties of the phenomena involved. Even Hartshorne, who has wrongly become the symbol of the anti-scientific in geography, recognized the danger of compartmentalization of the discipline when he wrote in 1959 that 'the division of each half (physical and human) into sectors based on the similarity of the dominant phenomena is of relatively recent origin and has proven detrimental to the purpose of geography' (146, p. 79). Indeed, the concept of a unified science of geography, a geography of a form that is not modified by some adjective relating to non-spatial properties, has nearly been cast into oblivion by the proliferation of modified sub-disciplines and the growth of hybrid disciplines such as regional science and environmental studies.

A brief and perhaps oversimplified account of the evolution of contemporary geography may prove instructive. As noted in chapter 1, in the period stretching from the late eighteenth century to the early twentieth century geography was both conceptualized and practiced as a unified science. Humboldt, Ritter, Kant, Ratzel, Hettner, Mackinder, Taylor, and Huntington, to name just a few, although holding a wide variety of views concerning the substance and methodology of geography, were in agreement that the discipline possessed both a discernible theoretical unity and the characteristics of a science.[13] The zeal of the attempt of the 'environmental determinist' school of the early twentieth century to establish broad and meaningful laws of human society led to a severe reaction within the discipline; the nomothetic, law-seeking endeavors of science fell into disfavor along with the sometimes bizarre (from our present perspective) laws postulated by the determinists. Many geographers came to support the notion that the development of meaningful generalizations about man and his relationship to the earth were not possible, and attention was turned to the study of the unique, rather than the general, aspects of the earth as inhabited by man; areal differentiation and landscape studies became the vogue.[14]

Eventually, in the post-World War II era, there began a counterreaction to the lack of rigor and of a scientific approach that had existed in geography since the environmental determinism debacle. It became apparent to some geographers that the discipline was not progressing in step with other organized systems of knowledge which had adopted an orientation toward analytical and empirical methods. In an attempt to regain respectability by adhering to the principles and methods of science, geography embarked on the 'quantitative revolution'. The rise (and, some would say, the subsequent decline) of the emphasis on the quantitative method is now a familiar tale. In retrospect, we may observe that the 'revolution' was the result of a genuine need to bring a more scientific methodology into geography, of a concern to develop a body of theory, and of a dissatisfaction with the idiographic form of the discipline.

Contemporary geography, then, is a field characterized by diversity, rather than unity, and specialization, rather than generalized coherence. The discipline, as engaged in today, is the end product of a complex series of multiple fragmentations. Among these we may identify, first, the human–physical dichotomy; second, the tendency to create special fields as the result of the stress placed upon the non-spatial properties of phenomena under investigation; and third, the emphasis upon techniques attendant on the rise of quantitative methodologies. Thus, we now must contend not only with a proliferation of sub-disciplines but also with an emphasis upon techniques of measurement and testing within each of these sub-fields. Fourth, there is the distinction between quantitative and non-quantitative methodologies; this may even rival the human–physical division in its magnitude. And, finally, there is a dichotomy of structure versus process. Although, as noted, Bunge speaks of structure and process as the complementary duals of geography, it has been argued that such a conceptualization is inadequate.

This divisiveness is readily observable in the structure of the curriculum and of the literature of the discipline. The human-physical dichotomy is perhaps the most obvious aspect of the fragmentation as the vast majority of geographers, beyond the first year of their university education, identify themselves with either the human or the physical branch of the discipline. This is no doubt a function of the pre-programmed 'tracks' through the course offerings of most departments. In addition, there are a number of geographic journals that address themselves exclusively to one or the other branches of the discipline, the recent *Progress in Human Geography* and *Progress in Physical Geography*, and the A and B Series of *Geografiska Annaler* being notable examples.

For evidence of the creation of specialized *geographies* based upon the non-spatial properties of phenomena one need look no further than the nearest university course catalog. There one finds courses on the cultural, economic, demographic, agricultural, pedological, and climatological aspects of spatial distributions. The distinction between quantitative and non-quantitative approaches is manifest in both course offerings and the journals. With few exceptions, the articles published in *Geographical Analysis* or *Economic Geography* are difficult for an individual lacking a fair amount of training in quantitative methods to comprehend. In addition, some of the papers deal with techniques so specialized that one would not expect them to be fully understood outside of a small group of the initiated. On the other hand, articles requiring quantitative training for comprehension are, in general, a rarity in a second class of journals, characterized by *The Geographical Journal*. There is, of course, a third class of journals which attempt to present a balance of the two approaches.

In relation to the above issues, the dichotomy between form and process approaches may seem to be of minor significance. This distinction, however, forms one of the major points of contention over the methodology of the discipline. Schaefer (264), Sack (259; 260; 261; 262), Eichenbaum and Gale (100), Berry (25), Harvey (147), and Bunge (55), among others, have expressed views on the role of these concepts in geographic inquiry.

While a measure of diversity is probably a healthy sign in any discipline, divisiveness probably is not. Within a fragmented discipline, one lacking a distinguishable conceptual framework, there may be the danger that individual sub-fields will become isolated from one another and will maintain little intellectual intercourse. The arguments against 'patchwork' approaches in science were outlined above in the context of general systems theory. These arguments are reaffirmed here. A conceptually unified geography may be a more effective vehicle for seeking knowledge about man and his environment than a number of loosely related *geographies.*

A methodological basis for overcoming the effects of the fragmentations that have been identified in the discipline may exist in one or more of the systems approaches to geography. These, in turn, are closely related to theoretical geography. In the remaining portion of this chapter the characteristics of theoretical geography are examined. This commences with a brief summary of the role of theory in science.

IV. The Role of Theory

In chapter 1, theoretical science was examined and found to consist of logically and economically ordered knowledge, a catalog of known propositions (as distinct from facts) in a rational order. Theory was the identified medium through which the foundation of science, consisting of economy of thought, conceptual simplicity, and the efficient organization of propositions, may be achieved. In the present section, this view of the role of theory in science is expanded, with particular emphasis upon the role of theory in geography.

Bunge (55, p. 2) observes that science is composed of three elements: logic, observable fact, and theory (Figure 2.1). A theory is formed by the union of a logical system (dealing with the relations between symbols) with operationally defined facts. Theory, he continues, is the heart of science because scientific theory is a key to the puzzles of reality. Observable fact or logic alone cannot begin to penetrate the essence of the real world. The creation of theory is difficult because the scientist must successfully identify the purely logical symbols of mathematics with a set of observable facts. Then, once theory is produced, it can often be applied with success to a variety of subjects.[15] In this sense, there is a unity of knowledge.

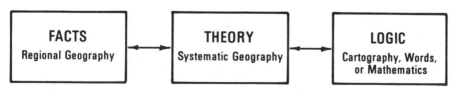

Figure 2.1 *The interaction of fact, theory, and logic in geography*
  Source: Bunge (55, p. 37)

Wilson (360, p. 32) defines theory as a set of propositions that seeks to explain the structure of some system and/or how the system develops. Thus, theory may involve a consideration of pattern or process, or both. In Wilson's view, the key term in the definition of theory is 'explain'. In the case of some complicated systems, however, explanation can only be achieved by the employment of some form of mathematical language.

Theory must extend beyond that which is directly observed. Its principal task is to uncover the objectively existing relationships that determine the essence of the phenomena being studied. By creating a general idea about the object and by revealing the relations that characterize it, theory suggests the course of further study by uniting the efforts of those who are concerned with the study of individual elements (9, p. 72). Theory is thus the principal means of ensuring that we will not lose sight of the essence of the whole because of its differentiated analysis. Yet, by no means can the role of theory be reduced merely to that of generalization of factual material. The essence of an object of study can never be understood in the process of direct inquiry alone. Only phenomena can be perceived and investigated directly. In order to proceed from the cognition of phenomena to the cognition of essence, theory is indispensable (10, p. 123).

A theory, then, must be an abstract method differing from that of mere registration of the results of direct observation and experiment. Theory must go much further than experiment in providing a new, more profound and broader generalization of knowledge expressing the active penetration of human thought into an objective material reality. Anuchin (11, pp. 44-5) tells us that theory must be able to give a fuller understanding of science as a whole, and must be able to supply the scientist with the logic of his discipline. He continues that the logic of any science lies in the definition of its substance and of its methodological basis. Such a logic can only form a really sound theoretical conception when it is based on adequate definition of the subject matter with science (11, p. 45).

We have seen that the creation of theory involves abstraction. Many authors, among them Anuchin (11), agree that one way out of the current crisis of the empirical domination of science lies in the strengthening of abstract thinking. This will result in the fusion of scattered information and conclusions of specialist investigations into rigorous and elegant logical systems and make possible the acceptance of empirical results as the content of a legitimate discipline (11, p. 44). In other words, only through abstract thought is it possible to synthesize the results of specialist observations and, more importantly, to reach the level of generalization that is the hallmark of science. Georgescu-Roegen (113, p. 319) observes that abstraction is the most valuable ladder in any science, and cites Marx's argument that abstraction is all the more indispensable in the social sciences since there 'the force of abstraction' must compensate for the inability to employ measurement tools or experiment.[16] Mathematics has emerged as one useful form of abstraction. Its use enhances the possibilities of gaining knowledge about real life and enables us to discover

46

a formal unity in diversity and to establish general regularities without being diverted by a host of qualitative differences (9, p. 76).[17]

V. Theory and Geography

When the role of abstraction is recognized and theory is employed in a meaningful way within the discipline of geography, the result may be deemed theoretical geography. One of the first geographers of the modern era to identify and stress the need for the utilization of theory in the discipline was William Morris Davis (88), who recognized that a scheme of geographical classification founded on structure, process, and time (as was his own work on the landform cycle) must be deductive in a high degree.

> As a consequence, the scheme gains a very 'theoretical' flavor that is not relished by some geographers, whose work implies that geography, unlike all other sciences, should be developed by the use of certain of the mental faculties only – chiefly observation, description, and generalization. But nothing seems to me clearer than that geography has already suffered too long from the disuse of imagination, invention, deduction, and the various other mental faculties that contribute towards the attainment of a well-tested explanation. It is like walking on one foot, or looking with one eye, to exclude from geography the 'theoretical' half of the brain power, upon which other sciences call as well as upon the 'practical' half. Indeed, it is only as a result of misunderstanding that an antipathy is implied between theory and practice; for in geography, as in all sound scientific work, the two advance most amiably and effectively together. Surely the fullest development of geography will not be reached until all the mental faculties that are in any way pertinent to its cultivation are well trained and exercised in geographical investigation.

The role of theory was also identified by Sten de Geer (89 , pp. 1-2) in the early part of this century. He observed that it was not until facts began to be systematically brought into relation with one another through the use of theory, and conclusions to be drawn therefrom, that geography became a true science. Anuchin (11, p. 66), too, recognizes the role of theory when he suggests that the absence of a sound theoretical basis in contemporary geography is one of the chief reasons for the inability of geographers to answer questions that arise in practice. Scientific theory must go in advance of practice, guiding its development.

Theoretical or 'theoretical–predictive' (348, p. 9) geography is not an arbitrary division of the discipline in the manner of 'human' or 'physical' or 'economic' geography. Rather, theoretical geography represents the most philosophically and scientifically well-developed level of the discipline. Warntz (347, p. 2) points out that geographic inquiry can take place at three levels: the descriptive, the classificatory, and the theoretical–predictive. In order to comprehend fully the implications of theoretical geography, we must examine it within the context of the two subordinate levels, so that we may see the manner in which it is the culmination and logical extension of them.

The first level, descriptive geography, has been the primary concern of the discipline for much of its existence. A good deal of the rote learning, the antithesis of understanding, that has recently passed as geography in our schools is of the form of low-level description. On the other hand, accurate description is no mean endeavor. Verbal description of the earth, perhaps formally commencing with Greek chorology, has played an important role in our cognition of reality. Similarly, many of the sophisticated quantitative methods employed in modern geography are descriptive techniques, the impetus for the development of which lay in the recognition that the ability to describe accurately a phenomenon is a prerequisite to comprehensive understanding.

The second level of geography is the classificatory. Classification involves the grouping of elements with certain similar properties and is the basis of regional geography, which classifies the phenomena on the surface of the earth. Unlike description, classification in geography involves the notion of location. At a sophisticated level, classification can be viewed, as Warntz points out, as an extension of point-set theory in its full geometrical implications to the geography of the earth's surface, which is the 'set of all sets.' Bunge (55, p. 234) tells us that classificatory geography answers the question 'What is where?', but this leaves geography lacking in explanatory or predictive power. It is the role of the highest level of geography to consider the question 'Why the where?'.

The third and most advanced level of geography is the theoretical–predictive form. It is the most recently developed of the three and subsumes the approaches of the lower two. Theoretical–predictive geography has emerged as geographers have begun to recognize the relationship and common characteristics of form and movement. Geographical prediction or the explanation of the location of either human or physical phenomena involves reference to movement. Human and physical phenomena are not static, and in order to 'understand' them we must seek to explain why and how they arrived at their current locations, and how their movement is likely to alter their pattern in the future. Movement produces the changes in the spatial arrangement of features on the earth's surface, the primary concern of geographers. There is a tendency for interacting objects to arrange themselves as near to each other as possible (movement). In Bunge's (55, p. 235) terminology, this is the nearness problem and is the central theoretical problem or theme of geography. Theoretical–predictive geography recognizes the relationship between structure and movement as coequal approaches, both of which must be taken into consideration. It is this duality which differentiates theoretical geography from the realm of geometry. Theoretical geography, then, refers to that level of geography, the foundation of which is in deductive–nomological theory and the goal of which is explanation. Deductive–nomological theory seeks an understanding of broad classes of phenomena and requires the viewpoint of a heightened level of abstraction; explanation is the highest level of knowledge, implying a thorough understanding of the processes through which phenomena manifest themselves.[18]

The important role of abstraction in the development of a science was noted above. It comes as no surprise that abstraction commands an equivalent position in geography. Theoretical geography engages in a search for lawful spatial order. This quest becomes feasible only at an abstract level, since the uniqueness of each part of the earth's surface is all too evident. General regularities and systematic relationships can only be discovered through a consideration of the general qualities of circumstances in material objects. It is the general spatial properties of phenomena – situation, extent, form, and distribution – rather than the specific non-spatial qualities, the study of which belongs to the specialist disciplines, that forms the substance of theoretical geography. The basis of the discipline's methodology is the progression from the uniqueness of statements of fact to the level of general proposition that provides the desired integration which characterizes all theoretical sciences.

Bunge (52, p. 323) argues that the basic language of theoretical geography concerns abstract relationships, Schaefer's 'spatial relations', rather than the non-spatial specific qualities of the objects under consideration. This implies a unity between human and physical geography, and, further, between all of the various systematic branches of the discipline. This unity is manifest at the theoretical level where common spatial properties and dimensional similarities can be observed. Warntz (348, p. 10) reinforces this argument by noting that geographers have been able to generalize their work concerning the phenomena of spatial patterns of human activities to such a level as to show these patterns to be isomorphic with the kinds of spatial patterns exhibited in physical geography as well. There is emerging, he observes, a unified theoretical geography which, through mathematics (principally the various geometries and topologies) and cartography, studies, explains, and demonstrates the significance of spatial relations in general, regardless of the vast non-spatial differences that one may recognize among phenomena. Tschierske (321, p. 106), too, states explicitly that formal theoretical geography is able to cut across traditional geographical distinctions. His definition of 'formal' requires that such laws will have validity over all subject areas, even those which have traditionally been beyond the scope of geography because of the schism and autonomy of the 'daughter sciences.'

Before concluding this discussion it is necessary to point out that there is some ambiguity regarding theoretical geography. The confusion in geography between the terms 'theoretical' and 'quantitative', in particular, has been noted by Wilson (360, p. 31). This view is affirmed by Taylor (304, p. 137), who points out that most of the research in the 'new' geography has not searched for *general theory*, for axioms or, in fact, for theories of any sort. Rather, he continues, the emphasis has been empirical, involving the use of various quantitative techniques on areal data of various types. 'Thus the terms "quantitative geography" and "spatial analysis" are accurate descriptions of the majority of recent work in geography.'

For direct evidence of this confusion in terminology one might cite *Geographical Analysis,* a journal subtitled 'An International Journal of Theoretical Geography.' While *Geographical Analysis* does in fact contain articles that may be termed theoretical, a majority of them are empirical, involving hypothesis testing (which is the mechanism for falsifying and/or refining theoretical formulations) or the development of algorithms for solving particular classes of problems. Using Taylor's terminology, the journal may be more correctly regarded as one primarily concerned with advances in spatial analysis.

This distinction may be largely one of semantics. The intent of the above discussion is not to imply that spatial analysis or quantitative geography, which often involve a calculus by which primitive terms and axiomatic statements are used to formulate *theorems,* are not directly related to the construction of theory in geography. One might argue, however, that spatial analysis does not provide a 'nesting place' (7, p. 63) for regularities that have been observed, a framework for the efficient organization of facts. Theory does not necessarily have to be quantitative or mathematical (360, p. 32); the set of propositions which compose theory may be expressed in a variety of other languages as well. The theoretical geographers that have been cited have generally employed mathematics, but in such a way that it supplements their discussions of the efficient organization of, and need for *general theory* in, the discipline.

The status of theory in geography is at present largely prescriptive rather than descriptive of the 'state-of-the-art'. That is to say, theory, in the formal sense of a body of general laws or principles that explain or account for phenomena or processes, is primarily a goal to be sought after rather than a commonplace of geographic experience. The 'theory' in theoretical geography at present may be regarded as a less formal set of principles that enables one to systematize economically and generalize the relationships between individual facts. This less formal body of theory, which falls between the formal usage and the loose sense of speculation or conjecture, may be viewed as a vehicle through which formal theory is pursued.

A final instructive point of a more general nature is suggested by Wilson's (360) brief review of the development of theoretical geography. His list of individuals who have contributed to the development of theory in geography contains a majority of non-geographers, encompassing the fields of economics, sociology, regional science, physics and geology. This would seem to have important implications regarding the trans-disciplinary spatial science approach discussed above.

A number of authors (360; 82; 225; 296) have explicitly noted the parallel development between 'systems analysis' or 'general systems theory' and modern quantitative–theoretical geography. Chapter 3 amplifies upon this parallelism by examining attempts to introduce 'systems thinking' into geography. As will be seen, these attempts were, more often than not, undertaken at a conceptual level. The operationalization of systems approaches in research methodology will also be examined in the next chapter.

Notes

[1] For example, Thom (306), Wolpert (372), Elsdale (103).

[2] The term 'space-occupying processes and events' is, in fact, redundant. Each process or event that may be identified necessarily occurs in time and in space. As process is sometimes associated with only the temporal dimension, this usage has been purposefully employed here in order to make explicit the spatial aspect of process. (See chapter 1.)

[3] See, for example, Hall (142), Sommer (282), Lowenthal (192), Lynch (196), Piaget (235), and Gould (126).

[4] It has been argued here that unique phenomena can exist when only non-spatial properties are considered. This raises the question of whether it is possible to identify a process or phenomenon as 'uniquely geometrical.' The geometrical concept 'point', having position (location) but not magnitude or extent, has no non-spatial properties. Is, then, a point uniquely geometrical? It is argued that uniquely geometrical phenomena do exist but only in a dialectical sense (see p. 32 n. 13). These uniquely geometrical phenomena are abstractions and derive their uniqueness by virtue of their possession of no properties that need to be negated in order to consider them in general terms. That is to say, their uniqueness lies in being non-unique; their speciality derives from their generality. In the abstract, objects and events can have only spatial properties, while in the operational real world objects and events must have both spatial and non-spatial properties. All objects and events have both abstract and operationally concrete aspects. Just as the physicist finds it useful to regard the electron as a wave in some instances and as a corpuscle in others, the geographer must be prepared to deal with processes and phenomena both in abstraction and in specificity, as the situation warrants. The statement that there are no uniquely geographic (spatial) phenomena is made in the context of the specificity of the non-spatial phenomena being discussed in this paragraph.

[5] Interaction studies that demonstrate that space or, rather, distance – in absolute terms – can influence not only location but also non-spatial properties include Zipf (379), Stewart (291; 292); Warntz (334; 336; 342), MacKay (199), and Olsson (227).

[6] A dissenting view is held by Sack (260; 262) and others, who regard space or distance as a surrogate for cost, time, energy, and so forth. See Pooler (241) for a discussion of the emergence in geography of the view that space plays a causal role.

[7] See Warntz (342).

[8] Lukerman (194; 195) has also made criticisms but these are incorporated in the arguments of Sack.

[9] For example, Warntz (334; 336; 342), Olsson (227), Stewart (291; 292), and MacKay (199).

[10] The somewhat absolutist view of space and distance that has emerged in this section is balanced in chapter 4 by a discussion of various ways in which distance may be defined.

[11] Here Sack ignores the 'inference problem' discussed by Harvey (148), Olsson (226), and others, namely that any number of processes can be inferred from a given form.

[12] Wolpert (372) notes that a useful distinction must be made between pattern and

form, two terms that are often applied interchangeably as synonyms of spatial structure. He points out that form involves movement and change in shape and location, whereas pattern does not involve change or movement but merely the specification of spatial differences.

13 Anuchin (10, chapters 1-3) contains a good summary.

14 See Sauer (263) and Hartshorne (146).

15 Gunnar Olsson conceptualizes the role of theory in this context:

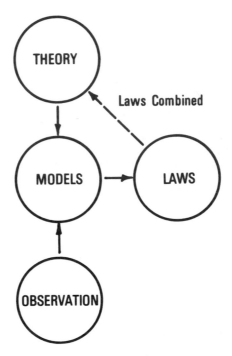

Figure 2.2 *The role of theory*

Models are the means by which our theories about reality and our observations of reality are joined. The successful combination of theory and observation, through the model, leads to laws which, combined with other laws, enable us to form new theories. This 'model of models' was brought to my attention by Jim Pooler, a student of Gunnar Olsson.

16 Karl Marx, *Das Capital* (3 vols, Chicago, 1932-3), I, 12, quoted in Georgescu-Roegen (113, p. 319).

17 The above brief treatment of theory should perhaps be supplemented with a more explicit exposition of its logical structure. For an excellent discussion of theory, the reader is referred to Harvey (147, Chapter 7).

18 Warntz (348, p. 9) and Bunge (55, p. 7) imply that prediction is the highest level of knowledge. Strictly speaking, this is not entirely accurate. The ability to explain includes the ability to predict; the ability to predict, however, does not necessarily include the ability to explain. For example, one may use regression analysis to predict the value of the dependent variable when the independent variable is at a particular value, although one is not able to *explain* the nature of the relationship between the variables.

3

Systems Approaches in Geography

An appropriate complement to the preceding chapter on the nature and methodology of geography and an essential component of an examination of systems frameworks for the discipline is a review of the antecedents of geographical 'systems thinking' and some of its earliest manifestations. Due to the volume of literature that has involved some form of 'systems approach', no attempt to be exhaustive is made; rather, the intent is to survey a number of studies that have exerted an influence upon the discipline. For the sake of clarity, in the following discussion the terminology set out in chapter 1 is employed, rather than that which may have been utilized by a particular author. In the first section, explicit attempts to introduce systems thinking into geography are examined; these attempts were, in general, at the conceptual level. A number of implicit but operationally oriented examples of the systems approach in geography are then considered. As will be seen, some of these predate the explicit concern with systems in the discipline.

I. Explicit Systems Thinking

The discipline of geography does not exist or function in isolation; as is the case in most other fields, its methodological and philosophical positions are closely related to the general intellectual climate of the day. The classical and regional approaches to geography, for example, evidenced certain commonalities with Darwinian and social anthropological schools of thought, respectively. In a similar way, both theoretical geography and the 'systems approach' to the discipline were related to a general interest by post-World War II science in the conceptualization of objects and processes as systems. This concern with systems, in turn, can be traced to a number of sources. First, there developed an interest in and capability with multivariate statistics, linear programming, input-output analysis, operations research, and similar research techniques designed to further the understanding of functional interdependence between objects belonging to the same set (225). Second, there was the rise of General Systems Theory. Such writers as Bertalanffy (30) and Boulding (40; 41) argued that the interdependence and functional relationships within a system, its organization and interaction, should in particular be studied. From the 1950s onwards model building was being simultaneously developed in many disciplines, and it began to be possible to perceive features of a general structure among all the system models which were materializing.

Curry (82) has identified one pragmatic utility in the type of general theory after which GST seeks: two or more sets of events not having any conceivable dependence can nevertheless be thought of as obeying the same 'laws'. Complex processes may have analogues in different phenomenological fields with common or similar methods of mathematical formalizing. 'The possibility of the intellectual economy offered by such a theory is its justification.' He continues that, at the formal level, it is advantageous to have a language divorced as far as possible from any phenomenological content. It is the quest for this type of theory which has fathered both the 'systems approach' and theoretical geography; in each case a neutral vocabulary for discussing disparate phenomena is sought. Curry concludes that the possibility of having this type of theory available in geography is appealing, 'burdened as we are with such a wide range of phenomena.'

Although the formal application of systems concepts has been a relatively recent development in geography, the notion of a system is not a novelty to the discipline. The history of systems thinking in geography is closely related to the functional approach, the concept of regions as complex interrelated wholes, and to the ecological approach to the discipline (147, p. 467). 'Systems thinking' may be identified in the writing of de la Blache, Ritter, Sauer, Barrows, and others. To a great extent, however, the notion of a system was implicit in and peripheral to the work of these authors.

A. INITIAL STATEMENTS

During the 1960s a considerable number of works were published in which the utility and the appropriate methodology of a systems approach in geography were explicitly treated. The formal introduction of the 'systems approach' into the discipline was signalled by the statements of Blaut (36) and Chorley (68) in 1962, followed by that of Ackerman (4) in 1963. In a methodological paper, Blaut stressed the need for geographers to abandon the notion of natural entities as objects of study, and for them to concern themselves with 'sets of interlocking propositions about systems.' The areal integration of systems was regarded by Blaut as the core of geographic endeavor.

Shortly thereafter, Ackerman presented the criticism that many geographers have tended to view geography in isolation, as an end in itself, rather than in the broader context of a contributor to a larger scientific goal. Both the development of a professional identity and the attainment of further progress within the discipline were, he argued, contingent upon a reappraisal of the overriding problem recognized by geography. Areal differentiation, long regarded as the primary focus of the discipline, was no longer a suitable candidate for this role as it did not lead to a common basis with other sciences. Identifying the 'systems approach' as one of the major methods in problem solving – ranking with the development of writing, arabic numerals, analytic geometry, and calculus – Ackerman redefined the central concern of geography as an understanding of the 'vast, interacting system comprising all humanity and its natural environment on the surface of the earth.'

In Ackerman's judgment, the use of the systems concept has three advantages. First, it allows us to view a set of problems at different levels in a way that has never been available before; second, it makes available to us a number of 'sharp new problem-solving methods'; and, third, if we are willing, it also places us in association and in close communication with other sciences whose overriding problems are similar. The techniques of systems analysis are seen by Ackerman as of particular value to geographers in applying their organizing concept of space to the examination of sub-systems of the world-wide man–environment system. No attempt is made, however, to identify these techniques or to outline their possible utility.

In contrast to these two very general arguments in favor of a systems approach, the paper by Chorley is marked by greater specificity; it represents the first attempt in the discipline to provide an explicit framework for geographic inquiry through the application of systems concepts.[1] Much of the paper is devoted to an interpretation of the erosion cycle formulation of William Morris Davis in systems terms. The Davisian scheme is criticized on the basis of being a 'closed system' approach to open systems of landscape. The distinction between open and closed systems is established in some detail and the value of an open systems view in geomorphology is made explicit. In addition, a number of other systems concepts such as equifinality and feedback are introduced and their utility outlined. Chorley judges systems theory to be of considerable value in geomorphology as it may increase the scope of the study, make possible correlations and associations which would otherwise be impossible, and generally liberalize the whole approach to the subject. Further, it allows geomorphology to be integrated into a wider conceptual framework.

B. CONCEPTUAL AND SUBSTANTIVE WORK

Due to the impetus provided by these three seminal papers, the 'systems approach' quickly gained the attention of geographic researchers. In the resulting 'second wave' of systems literature several specific areas of emphasis may be distinguished. One set of work is largely an extension of the viewpoint presented by Blaut, Ackerman, and Chorley; it is characterized by the reaffirmation of the utility of the systems framework in geography. In his discussion of the substantive and methodological aspects of geography, Berry (20) underscores the need to recognize the interrelatedness of various systems that exist on the earth's surface. The integration of concepts and the search for generalization that characterize GST are suggested as a viable method of achieving a synthesis of various approaches to geographic understanding. Noting that geography's integrating concepts and processes concern the world-wide ecosystem of which man is the dominant part, Berry proposes the systems perspective as a suitable framework for the analysis of regions.

In 1965 the National Academy of Science released a report entitled *The Science of Geography* (218).[2] Within the general context of a discussion of the role of geography in modern scientific research, two aspects of systems thinking within the discipline were stressed. First, the primary concern of geography was defined as the man–environment system. Although geography shares this concern with other disciplines, it studies it from the viewpoint of 'space in time'. Further, the systems concept has opened the way toward a flexibility in research that is freeing the field from a past view of the world as a mosaic of regions. Second, the need for the development of a common language for facilitating communication between geography and other branches of science was emphasized. Systems techniques and concepts would play a central role in such a language, it was suggested.

Making direct reference to Chorley's work in geomorphology, Haggett (140) advocates regarding regions as open systems. The benefits of such an approach are identified as, first, the ability to direct our attention toward the links between process and form; and, second, the integration of geography with the other sciences that organize their thinking in this manner. In two papers on the ecosystem concept, Stoddart (295; 296) amplifies upon these points. He notes that, by adopting a systems approach, geography 'no longer stands apart from the main stream of scientific progress.' Not only does the systems conceptualization have a high degree of applicability and enable researchers to pose new problems, but, through its emphasis on organization, structure, and functional dynamics, it also provides the discipline with a methodology. Both Stoddart and Kates (167) point out that the systems approach is monistic; it unifies the disparate elements of the man–environment system in a single framework within which the interaction between components can be analyzed. The conventionally distinct treatment of the human and the physical sub-fields of the discipline may thus be seen to lack a logical basis.

In addition to making the customary exhortation for readers to familiarize themselves with the many benefits of adopting a systems approach, Wilbanks and Symanski (358) make the first rigorous attempt to eliminate some of the ambiguity in systems thinking by distinguishing between 'substantive' and 'methodological' approaches to the study of systems. The former is based upon the premise that 'useful research can only be done by breaking a complex problem or entity into individual components which are studied one at a time.' On the other hand, the latter involves the use of 'formal, axiomatic means for organizing and analyzing knowledge in complex pieces of research.' Although this distinction is a useful one, the contribution of Wilbanks and Symanski is somewhat diminished in that it does not adequately distinguish between systems analytic and General Systems theoretic endeavor.

The preceding papers are very general conceptual discussions of systems thinking. A theme common to all of them is that the systems approach provides geography with a useful methodology for both viewing the world and relating to the other sciences. A second set of work is characterized by an increased concern with the properties of systems. Three papers adopt a true General Systems approach, emphasizing the isomorphism between diverse phenomena. Woldenberg and Berry (369) outline the homologies in the behavior and attributes of cities and rivers, both of which are identified as open systems exhibiting dynamic equilibrium, negative entropic tendencies, and equifinality. In addition, both classes of system exhibit allometric growth (see chapter 7). Berry (21) stresses the isomorphy between urban models and models in other types of enquiry, noting that urban theory contributes to the development of GST. The suggestion is made that cities and functionally linked sets of cities are systems susceptible to the same kinds of analysis as other systems and are characterized by the same types of generalizations, constructs, and models. Information theory is identified by Berry as the core of the systems approach. In a less explicit fashion, Curry (82) indicates that certain 'laws' and relationships which may be derived from the study of a central place system have analogues in other fields. The central place arrangement is discussed using a 'neutral' and broadly applicable language, consisting of terms such as filters, signals, noise, distortion, and spectra.

In contrast with these papers, Ajo (6) adopts a systems analytic approach in his attempt to model the behavior of a demographical system. Particular emphasis is given to the determination of the response of a system to a given excitation. This approach is similar to that outlined by Blalock and Blalock (35) in their seminal paper on systems analysis in the social sciences. Both Ajo and the Blalocks stress the functioning of systems, with particular emphasis on forcings (input) and responses (output).

Other substantive attempts to conceptualize research problems in systems terms include the work of Foote and Greer-Wootten (109) on modelling a simple system of cultural–physical–biological interaction in a northern environment, and that of Walmsley (332) in the human geographical contexts of learning theory and consumer behavior. One of the most detailed applications of the systems approach to a real problem is presented by Mabogunje (197). Rural–urban migration is formulated as a spatial process whose dynamics and spatial impacts must form part of any comprehensive understanding of the phenomenon. Further, the isomorphy between migration and other phenomena is made explicit.

The systems approach has also been usefully employed in a pedagogic context. This application of systems thinking was pioneered by McDaniel and Eliot Hurst (198), who utilized it to structure a course on economic geography. Besides introducing the basic systems concepts such as complexity and interconnectedness and formulating the spatial organiza-tion of the economy in such terms as the resource system, the tertiary system and so forth, there is, in addition, an explicit reference made to the role of systems analysis. The suggestion is made that the spatial structure of an economic system can somehow be manipulated so as to maximize some human welfare function. The approach outlined in this monograph is developed in some detail by Eliot Hurst (102) in an economic geography text. The most comprehensive pedagogical utilization of the systems approach in geography to date is the text, *Physical Geography*, by Chorley and Kennedy (69). This attempt not only recasts a wide body of existing information within a systems framework but also employs that framework to indicate the ways in which socio-economic and physical systems interlock and interact. More recently, Russwurm and Sommerville (258) have edited a volume of systems-oriented perspectives to man's natural environment.

C. DISSENTING VIEWS

A number of authors have presented dissenting views on the utility of the systems approach in geography; most of these authors, however, have exempted systems analysis from their criticisms. The most vigorous statement of dissent has been made by Chisholm (65) in a paper which concentrates primarily on Chorley's (68) analogy between the Davisian erosion cycle and a closed system. Although the systems analytic endeavor of devising more efficient ways of system functioning is not dismissed, the systems theoretic and GST approaches are regarded by Chisholm as 'incoherent strivings' (65, p. 48).[3] The vagueness of the systems concept and the shallow nature of the analogies employed by Chorley are criticized with some justification. The extrapolation of these criticisms to more recent and comprehensive systems research in the discipline, however, is not justified. Chisholm further asserts that the distinction between closed and open systems is a useless one because closed systems do not exist, and that systems theory is too concerned with inductive reasoning. The former point is, in itself, irrelevant as closed system models are sometimes employed in geographic research, while the latter overlooks the deductive portion of systems research, characterized by the work of Ashby and Mesarovic. In fairness to Chisholm, at the time of his criticism a great deal of substantive systems research had not yet been accomplished in the discipline. Commenting upon Chisholm's critical statement, Walmsley (332, pp. 36-7) suggests that he displays 'an emotional, rather than rational, reaction' since neither the claim that it is both premature and unnecessary to dignify the search for generality with a title nor the proposition that GST may be 'an irrelevant distraction' are supported.

Greer-Wootten (131) reviews some of the advantages and disadvantages of systems theory in geography and concludes that it cannot contribute to the development of geographic theory simply because this approach draws from all other sciences those common elements in their particular theoretical structures; specific instances of general concepts need to be clarified in a geographic context. He argues that if the systems theoretic approach has any relevance at all to the field it is in an indirect manner; it indicates that theory should be developed within the discipline, so that geography can ultimately contribute to systems theory. Especially when compared to systems analysis, systems theory has little direct relevance for the development of geographic theory.

Bunge (55, p. 251) takes a surprisingly narrow view in dismissing the systems theory/GST approaches in geography. Although he points out that, using his own arguments concerning the conservation of academic effort, the startling resemblances between geographic problems and those of biology might tempt one to conclude that a 'really efficient science of location' might include the spatial aspects of biology, physics and so forth as a study in general systems, he rejects this possibility. 'The very range of spatial scale between various sciences introduces worlds so weird that spatial experts in one have little to say to spatial experts in another.' This reaction is particularly puzzling, as in several places in the same work he takes pains to emphasize the relevance of the work of the biologist D'Arcy Thompson (309) to geography. It would seem that Bunge is ultimately bound by the traditional notions as to what constitutes geographic research.

D. CONCLUSION

Finally, spanning all aspects of the systems approach is the work of Harvey (147). In a chapter of his *Explanation in Geography* he presents a comprehensive summary of the various facets of systems thinking in geography. Although failing to distinguish clearly between systems analysis and the non-operational approaches, Harvey covers important definitions, typologies, criticisms, models, and methodologies concerned with the study of systems, relating these directly to geographic inquiry wherever possible.

At the time of this statement, 1968, Harvey correctly noted that the systems approach had not gone very much beyond the stage where we were exhorted to think in systems terms, and that the employment of systems concepts and thinking had not yet achieved a powerful operational status in the discipline. The concluding portion of this chapter attempts to point out some advances that have been made in the application of systems thinking since that time. Before this attempt, however, it will be demonstrated that implicit operational applications of the systems approach preceded the initial explicit conceptual statements by Blaut, Ackerman, and Chorley.

II. Implicit Systems Approaches

Just as the conceptualization of various objects and processes in the real world had been formulated in systems terms long before the development of the various explicit aspects of the study of systems, the implicit use of systems approaches in geography predated the explicit statements that have been reviewed. The principal manifestations of implicit systems thinking are central place theory, potential models, and dimensional analysis. Underlying each of these approaches to spatial analysis is the fundamental concern with the functional interrelations between the elements of a set of objects distributed in space and, moreover, with the extent to which relative location plays a role in the maintenance of those interrelations.

A. CENTRAL PLACE THEORY

As every geographer surely knows, central place theory is a static model of the distribution of a functionally differentiated set of settlements and their corresponding market areas. The foundation of the theory was laid in 1933 by Walter Christaller (70) in his examination of regularities in the settlement pattern of Bavaria. His attempt to explain deductively these regularities led to his postulating a settlement model with certain fixed qualities and assumptions.

The now familiar arrangement of a set of hexagonal market areas with their center-point settlements forming a triangular lattice results from the assumptions of (a) an unbounded isotropic plane with (b) an equally spaced, discontinuous population; (c) a differentiated set of goods and services which have varying thresholds and (d) which consumers purchase from the nearest central place; and (e) all parts of the plane being served by some central place.

Settlements are not only distributed over space but also differentiated in terms of a functional hierarchy of available goods and services. The functional or hierarchical characteristics of a settlement influence and, in turn, are influenced by its relative location in the set of settlements. Higher order places, offering more goods and services, having more establishments and functions, and having larger populations and market centers, tend to be more widely spaced in order to acquire a threshold level of market population. Lower order places, providing only low order or necessity goods requiring frequent purchasing and, therefore, being characterized by short, frequent consumer trips, have smaller market areas which may be in direct propinquity to one another. Since higher order settlements perform all of the functions not available in the lower order centers, a nesting pattern of lower order centers results. This hierarchy of settlements may be organized according to several principles, the most commonly employed of which are the market, transportation, and administrative principles (see chapter 8).

Christaller's formulation has been generalized by Losch (190), who demonstrated the manner in which a complete economic landscape can be created, based upon a general concept of hierarchies. Berry and Garrison (26) further generalized the model by demonstrating that a hierarchy is the direct consequence of the concepts of threshold and range, irrespective of how population is distributed. They also demonstrated that the model applies to the arrangement of business centers within cities. Subsequent to this work, numerous studies have refined, generalized, and tested the theory.[4] Perhaps the most recent development has been an attempt by White (351) to develop a dynamic central place theory.

Central place theory is perhaps the first systems formulation to develop specifically within the discipline of geography. Further, in contrast to the manifestations of the systems approach that have been examined above, the central place model represents an operational basis for implementing and testing the conceptual foundations on which it rests. That is to say, in the place of exhortations to think systemically, the central place model presented an opportunity to develop measures of spatial and functional centrality; to deduce and to test rules for estimating the size of a settlement and its market area from information concerning the configuration of the entire set of settlements; and to develop general principles concerning consumer travel behavior and the threshold and range of goods. In short, the theory was subject to empirical testing and refinement.

Essentially a model of relative location within a national, regional, or urban social–economic system, central place theory has implications for both the spacing and the hierarchical functional role of settlements. The location and the functional order of any one settlement are postulated to be directly related to the locations and orders of all other elements in the set of settlements. From a slightly different perspective, the model may be seen to represent a hierarchically organized distributional system which approaches, in its idealized form, a least-effort configuration. The original static model represents an equilibrium condition among the market centers; recent work (351) has investigated the results of causing disequilibrium in the system by adding or removing centers.

Finally, to place the central place model even more explicitly within a systems framework, there can be little doubt that Berry's extensive work (24; 26) on both inter- and intra-urban central place hierarchies has been a major influence upon his conceptualization of 'cities as systems within systems of cities' (21).

B. THE POTENTIAL MODEL

The antecedents of the potential model lie in the social science gravity model and, through it, in astrophysical science. The gravity concept of human interaction postulates that an attracting force of interaction between two areas of human activity is created by the population masses of the two areas and that a friction against interaction is caused by the intervening space over which the interaction must take place. This is, of course, a social application of Newton's (*Principia*, 1687) finding that the gravitational attraction between planets varies directly with mass and inversely with distance. The expectation in this 'social physics' is that interaction between two centers of population concentration varies directly with some function of the population size of the two centers and inversely with some function of the distance between them.

The earliest known explicit formulation of the gravity concept of human interaction was made by Carey (59) in the mid-1800s and independently reproduced in partial form by Ravenstein (247) in 1885 when he observed that migratory movement tends to be toward cities of large population and that the volume of movement decreases with distance between the source and the 'center of absorption.' The next development of the gravity concept appeared in the late 1920s when Young (376) made a similar attempt to measure the migration of farm population. At about the same time Reilly (250; 251) postulated his 'Law of Retail Gravitation' in order to examine the relationship between population size, distance, and retail sales. Using this technique he was able to compute the equilibrium point on a line joining two cities competing for retail trade.

In the early 1940s, John Q. Stewart (289; 290; 291; 292), an astrophysicist, and George Kingsley Zipf (378; 379), a philologist, independently generalized the gravity concept, verifying it on the basis of massive empirical evidence. It is these two researchers who are primarily responsible for the existence of the concept in modern social science; the concept has, however, received extensive development since their contributions.

The force of interaction between two concentrations of population acting along a line joining their centers is directly proportional to the product of the populations of the two centers and inversely proportional to the square of the distance between them:

$$F_{ij} = K\frac{P_i P_j}{D_{ij}^2}$$

where F_{ij} is the force of interaction between concentrations i and j, P_i and P_j represent the populations of the two centers, D_{ij} is the distance between the centers, and K is a gravitational constant. Following this physical analogy, the energy of interaction between the two centers, E_{ij}, which results from this force, would be:

$$E_{ij} = K\frac{P_i P_j}{D_{ij}}.$$

Thus, the energy of interaction between any two centers is directly proportional to the product of populations and inversely proportional to the distance separating them.

The gravity model of human interaction, like the Newtonian model of attraction between planets, is capable of dealing only with pairs of objects. That is, the relationship between only two population centers can be considered at one time. As with the astrophysical application of the model, it is possible to consider the relationship between all possible pairs in a set of centers in order to examine the net effects of the interactions occurring over the entire set of centers. This, however, is often difficult to operationalize and to interpret.

The elements in a set of population centers, in a manner like that of astrophysical bodies or electrons, exert their influence (force, energy, or attraction) upon one another simultaneously. That is, each center is simultaneously under the influence of *all* other centers and simultaneously exerts an influence upon *all* other centers, as well as upon *all points* located within the boundaries of the system. Thus, a field is created. The concept of a field originates in physics, where it was noted that in the manifestations of gravitational, magnetic, and electrical forces the bodies or charges involved exerted an influence not only at their point or area of location but also at a distance. This influence was shown to vary directly with size, that is mass or intensity, and to vary as the inverse of distance. The field quantity, which is the physical manifestation of this combined influence, can be measured at any point in a system, both at centers and in the intervening space.

John Q. Stewart (292) recognized the solution to the 'multibody' gravitational problem in human interaction; the solution was formulated as potential of population. This was a significant step in social science as it enabled a set of population centers, together with all points under their influence, to be treated by methods appropriate to the measurement of a system, in the true sense of a functionally integrated and interdependent whole.

The value of potential, V, at a point i is a definite integral:

$$V_i = \int (1/r) \, G dA,$$

where G is the density of population of an infinitesimal element of area dA, and r is the distance of this element to point i. This stems from the fact that the population distribution of an area is considered not as groups about points but as density distributions in a plane. The value of the integral cannot be computed directly, but a mechanical integration based on summations can yield approximations of the values. The procedure is a simple one:

$$V_i = \sum_{j=1}^{n} \frac{P_j}{r_{ij}} = \frac{P_1}{r_{i1}} + \frac{P_2}{r_{i2}} + \cdots + \frac{P_n}{r_{in}}.$$

The potential of population at point i is equal to the sum of the population at each point divided by the distance between that point and i. For convenience, the population is customarily divided into areal units, the centroids of which are used as the points between which the distances are measured and the populations of which are utilized in the above procedure. In addition, a value representing a unit's influence upon itself, where $r_{ij} = 0$, must be added into the sum V_i. There are several assumptions which have been used to estimate this value.

By plotting the potential of a number of 'control points' a map of the potential values over an entire system, a national social–economic system for example, may be constructed. Potential, a spatially continuous field quantity, has been demonstrated (293; 336; 337) to serve as a measure of position within a system, perhaps best conceptualized in terms of aggregate influence or accessibility. This is no mean achievement, for such a measure of position has been sought by geographers since the importance of the concept was recognized by von Thunen in the early nineteenth century, when he demonstrated that even if perfectly uniform land quality existed patterns of differential land use and differential economic rent would develop, based on relative position alone.

Stewart (290; 291), Stewart and Warntz (293), Warntz (336; 337; 341; 342) and others (77; 78; 94; 95) have demonstrated that geographical potential (a general term which includes population potential, income potential, market potential, and so forth) is associated with many social and economic phenomena, broadly classed as 'sociological intensity'. Among the areal correlates of geographical potential are land values, the incidence of stress related diseases, human migrations, and economic flows.

Figure 3.1 depicts the 1972 North American population potential surface. The lines of isopotential, in units of thousands of person per mile, depict the field of influence resulting from the force exerted by all pairs of units, in this case over 3500 US counties and Canadian census divisions. Using the values of potential as an aggregate index of influence or accessibility, one is able to estimate the influence exerted upon any given point by *all* persons in North America or, conversely, to estimate the relative accessibility at that point to *all* persons in North America. This map is considered to accurately represent the mutual influence of the elements of the North American social–economic system.

C. DIMENSIONAL ANALYSIS

Dimensional analysis receives a detailed treatment in chapter 5, where dimensionality is identified as a major component of spatial structure. This section does not attempt a comprehensive exposition of the concept but, rather, seeks to establish its identity as an early manifestation of the systems approach in geography.

The three physical units of mass, length, and time have become recognized as fundamental quantities; they can neither be derived from one another nor resolved into anything more basic. All other physical units are derived units and may be defined only by reference to one or more of these three units. Volume, for example, is derived from length in that it has the dimensions of length to the third power. Similarly, many geometrical properties which may be used to describe form are derivations of length. To these fundamental physical units may be added the fundamental social units of value and population size (293, p. 116). Population density has the dimensions of population divided by length squared, for example.[5]

Dimensional analysis is a tool which may be utilized after the dimensions of the phenomena under investigations have been identified. Corresponding to every physical or social quantity there may be ascribed a dimensional formula based upon the definition of the quantity in dimensional terms. The above definition of population density relates that concept to the set of fundamental quantities. The physical sciences have long recognized the principle of dimensional homogeneity. When the units

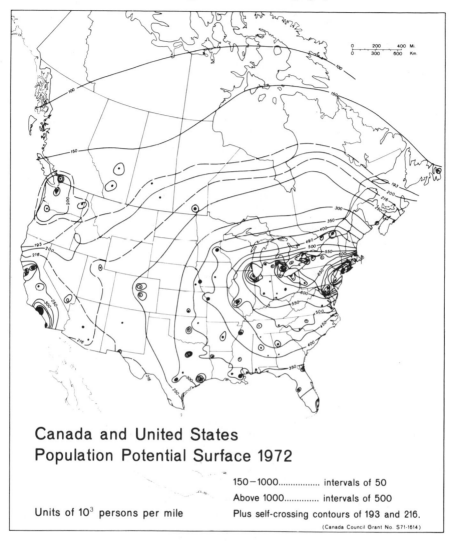

Canada and United States
Population Potential Surface 1972

Units of 10^3 persons per mile

150–1000................. intervals of 50
Above 1000............. intervals of 500
Plus self-crossing contours of 193 and 216.

(Canada Council Grant No. S71-1614)

Figure 3.1 *Canada and the United States: population potential surface 1972*

of an equation equate dimensionally as well as numerically the equation is said to be dimensionally homogeneous or balanced; the exponent of dimension which applies to any term must be the same as that in any other term. The technique of dimensional analysis may be used both empirically to verify quantitative relationships and deductively to formulate explanations and general laws through logical steps of reasoning from a series of initial postulates (297, p. 281).

Dimensional analysis was introduced from physics into the physical and the human branches of geography in the early 1950s by Strahler (297; preceded by a 1953 paper) and Stewart (290), respectively. An example taken from the work of the former will be employed to sketch briefly the relationship between dimensional analysis and a systems approach; the contribution of the latter will be left for consideration in chapter 5.

Strahler observes that, even if dimensional analysis served no other purpose than clarifying the essential nature of the elements measured, the technique would still be a valuable research tool. He then identifies two further steps that may be taken. The first of these is the formulation of equations through the Pi Theorem. In an analysis of the relationship between drainage density and six independent variables, Strahler employs dimensional analysis not only to set up a dimensionally correct basic form for the equation but also to derive dimensionless numbers (see chapter 5) which make explicit the functional relations which control drainage density.

The Pi Theorem indicates that the number of dimensionless or pi terms that result from an equation will be the difference between the number of variables and the number of fundamental dimensions employed. Since his equation utilizes seven variables which are derived from three dimensions, four dimensionless numbers may be expected. These Strahler identifies as the 'ruggedness', Horton, Reynolds, and Froude numbers. Drainage density is demonstrated to be inversely proportional to local relief times a function of the Horton, Reynolds, and Froude numbers. In addition, it is possible to reduce the original six independent variables to three variables directly related to these three dimensionless numbers. In this way, the complex set of interactions which produce a given drainage configuration are shown to be a direct function of a small number of readily measurable relationships.

In more explicit terms, a drainage network may be regarded as a highly interrelated system in the sense that an alteration in the relations between runoff intensity, relief, density of fluid transmitted, and so forth will alter other relations and, indeed, the morphology of the system. Dimensional analysis has been employed both to make explicit the nature of these relations and, therefore, of the overall systemic interdependencies, and to develop quantitative bases for the measurement of the relations.

The second application of dimensional analysis that Strahler identifies involves assessing the similarity of various physical systems. The similarity of two drainage basins, for example, is verified when two phases of similarity are demonstrated: (1) geometrical similarity in which the shapes correspond, and (2) dynamic similarity in which all forces in the systems are proportional (297, p. 291). On the basis of these criteria it is possible to evaluate the degree to which two drainage basins or two volcanoes are similar. In the case of drainage basins, if geometrical similarity exists the respective dimensionless 'geometry numbers', derived from the relations between the morphological characteristics of each basin, would be expected to be identical. In the same way, the dimensionless Horton numbers, indicative of the intensity of erosion processes, would be expected to be identical if the respective processes were similar. As may be readily observed, work of this nature is concerned not just with systems but with the concept of a general system which applies to a range of specific systems.

III. Recent Systems Approaches

In the years since Harvey's (147) observation that the systems approach had not progressed beyond general exhortations to think in systems terms, a growing number of geographers have utilized a systems framework, either explicitly or implicitly, in the analysis of complex problems. The nature of several of these approaches will be briefly reviewed. Once again, no attempt to be completely exhaustive will be made since, with the common acceptance in social science, as in science generally, of the existence of strong interdependencies between all aspects of man's physical and social environment, it is a rare piece of work which does not possess some recognizable elements of systems thinking. This section, then, considers some of the more obvious manifestations of a systems approach in recent geographical research.

A. THE DYNAMICS OF NODAL SYSTEMS

Along with central place theory, a classical approach to the examination of a set of nodes has been a graph theoretic framework. Graph theory has dealt almost exclusively with the development of structural models involving connectivity measures and path analysis, matrix theory, and the theory of radial structures (224;310). This highly productive area of research has recently been complemented by more direct attempts to analyze the dynamic conditional states of a model system. If a dynamic element is to be added to the conventional structural analysis it must include the consideration of the growth and decay of centers in response to changing patterns of spatial interaction.

The concept of an urban system implies that a change which occurs in any one city will affect other cities through a complex chain of reactions. One type of approach to the operation, rather than the structure, of a set of cities has been that of King, Casetti, and Jeffrey (171) and of Bannister (17). Recognizing that a network of highly interconnected places is characterized by interdependence, mutual causation, and lagged response, these studies have investigated the manner in which relatively stable model systems respond to short run changes in the form of economic impulses which may originate endogenously or exogenously and may vary in strength or timing. Methods have been developed for the determination of the patterns of interaction and for the identification of, first, which cities affect other cities and, second, the extent and timing of the effect of exogenous national factors upon each city.

A further contribution to this line of research has been recently made by Lamarche (175) within an explicit systems analytical framework. Lamarche employs flowgraph theory to measure the impact of network changes throughout an urban system. System gain, an interaction ratio of flows; transmission gain, an interactance multiplier coefficient; and system sensitivity, a measure of the susceptibility of the system to fluctuations, are used to examine the interconnection and operation of a set of interdependent nodes. System gain indicates the relative growth of the activity in dependent regions relative to that in the main region. Transmission gain describes the activity on the sub-networks linking one center to all other centers. System sensitivity reveals how each center reacts to changes in other nodes. The model is formulated as a set of simultaneous linear equations and holds promise for revealing insight into system operations that cannot be obtained from such techniques as regression and factor analysis.

A related concern with the operation of a set of nodes has been White's (351; 352) attempt to develop the framework for a dynamic central place theory. Noting that classical central place theory is largely unusable as it is based on rigid and unrealistic assumptions, White attempts to reconceptualize the theory in dynamic terms which are independent of those assumptions. The notion that the central place problem is essentially one of location is rejected and replaced by the basic premise that the problem is essentially one of examining the implications arising from the union of the microeconomic theory of the firm with the theory of consumer spatial interaction. The locations or potential locations of central places are regarded as given and fixed, and the problem is to describe the growth of each place in the system. The differential growth of centers will result in the appearance of a locational pattern of some sort, but questions of pattern are seen as of secondary interest. Questions concerning the behavior of the system are of primary significance. One important class of questions concerns the readjustment of the system following a shock to some part of it, for example the adjustment that would follow the establishment of major new industrial employment in one center.

White (352) postulates that the evolution of a central place system may be described by a set of first order difference equations and sets up two hypothetical systems, the first with one isolated activity and the second with two centers and one sector. Although these systems are simple ones, their behavior is considered to be suggestive of the kinds of relationships which may hold in more complex systems. In a later attempt (351), the necessity for examining the behavior of only simple systems, due to the impossibility of achieving solutions to large sets of simultaneous nonlinear difference equations, is overcome by use of simulation techniques; the implications of the theory for complex systems are explored.

Goodchild *et al.* (123) extend this approach by investigating the extent to which the form of a central place system, represented by the sizes and spacings of settlements, can be expected to be indicative of the processes occurring in the system. A model is constructed in which nodes grow or decline in response to migration, asymmetries in flows occurring as a result of temporary attractiveness advantages obtained by particular nodes.

B. SYSTEMS OPTIMIZATION METHODS

A second set of systems approaches involves the search for optimum solutions to clearly stated problems. This type of research may correctly be termed operations research or, in the terminology introduced in chapter 1, systems analysis. Operations research is a broad field which is primarily oriented toward business and economic topics. Many of the problems encountered in these fields have a spatial dimension and, therefore, are of inherent interest to geographers. Within a spatial context, these problems often involve influencing the structure or the behavior of a system by adjusting its spatial relations or the intensity and pattern of flows that occur within it. That is, the goal of the research generally involves the attainment of some sort of spatial efficiency.

The application of optimization techniques and, especially, the development of them is, in general, a relatively recent departure in geography because, as MacKinnon (200) points out, geographers have traditionally avoided normative frameworks, preferring instead to describe selected aspects of past, current, or future worlds 'unencumbered by any explicit goal orientation.' In geography, optimization techniques have been both applied and developed in the area of location–allocation analysis.

A location–allocation problem possesses two basic characteristics: a set of demand points (consumers) distributed spatially over an area and a set of supply points (facilities) to serve them. Subject to the exact specification of the location–allocation problem, the basic question is 'how can the supply points best be located with respect to demand and the demand best be allocated to the supply points?' As the name implies, the problem involves two interrelated stages, locating the supply and allocating the demand. The terms supply points and demand points are, of course, general ones which may be replaced by specific referents without altering the generality of the problem.

In the simplest case, one supply point is located with respect to a set of demand points. Since all consumers must necessarily be allocated to the single supplier, this is really a location problem and is solved by finding the point of minimum aggregate travel relative to the distribution of demand. The MAT point cannot be solved analytically, of course, and must be approximated through an iterative procedure. The problem becomes more complex as more supply points are involved.

The allocation problem, in its most elementary form, involves allocating demand to established supply points. The simplest method of allocation is to assign demand to the nearest supply point (the proximal solution), for example by constructing Thiessen polygons. This solution is often unrealistic, however, since there may be constraints upon the amount of demand that a supply point can handle. Schools have maximum seating capacities and, therefore, not every student is likely to be assigned to the school closest to his home. This general question has become known as the transportation problem.

The true location–allocation problem, then, involves the search for the best set of locations for supply facilities having constrained capacities and for the best allocation of demand to these locations. The optimum solution is that which minimizes an objective function, in many instances of the form

$$Z = \sum_{i=1}^{n} \sum_{j=1}^{m} C_{ij} X_{ij}$$

where C represents the costs of supplying one unit from each supply point to each demand point, and X represents supply capacities and demand requirements, the matrix of allocations. The minimization of the objective function is subject to the constraints that are imposed, usually that supply equals demand.[6]

Whereas other representations, the gravity model for example, attempt to describe what pattern of flows does or will exist, the location–allocation approach describes what patterns *should* exist in order to satisfy some criterion such as the minimization of aggregate consumer travel. Specific methods have been developed for both discrete (network) and continuous (planar) space. In addition, it is possible to utilize the approach as a benchmark for comparing with the existing structure of a system in order to derive a measure of spatial efficiency.

A second and closely related aspect of location–allocation analysis concerns districting or regionalizing, the delineation of the boundaries that implement a certain pattern of allocation. As seen above, when supply is unconstrained Thiessen polygons represent the boundaries which allocate demand to the nearest point. In cases where supply is constrained, districting is a matter of solving the transportation problem (375). Where the location of facilities (schools, fire stations, and so forth) is not specified, or where a large number of alternative sites is to be evaluated, an heuristic solution must be employed in place of a numerical–analytical solution.

An heuristic algorithm developed by Tornqvist (320) assigns demand to supply centers, thereby delineating catchment areas, when multiple supply centers exist. Such an algorithm will not necessarily produce the optimum solution but, rather, arrives at one which is near-optimal or convergent on a global optimum. Goodchild and Massam (124) use a converging or iterative algorithm to identify optimal administrative regions in Southern Ontario. After first defining eight regions corresponding to eight major cities, a process is carried out in which more efficient locations for the administrative centers of each region are found and a more efficient allocation of population to each center is made. There is, therefore, a continuing readjustment of both the facilities (administrative centers) and the boundaries of the regions which they serve. The algorithm finally converges on a 'stable' set of centers, the efficiency of which cannot be improved.[7]

A third related aspect of spatial optimization is the class of problems dealing with network flow and network structure. The simplest of the network problems is the so-called capacitated transportation problem, an extention of the more general transportation problem. Since the flows over a transportation network are restricted to the arcs of the network, the problem involves the carrying capacity of those arcs; flow must be allocated to the arcs subject to the capacity constraints. Other problems involve decisions as to how investments for improving a transportation network should be made while not exceeding budget constraints. The network improvement decision is whether or not two nodes should be linked and, if so, what the capacity of the link should be. As with the other classes of location–allocation problems, both numerical optimization procedures and heuristic solutions are available; the former can only be applied to discrete space formulations, while the latter must be applied to problems involving continuous space.

In the three related classes of systems optimization discussed above, those problems having a numerical optimization solution may, in general, be solved through linear programming methods. In addition, other forms of mathematical programming have direct applicability to this class of optimizing problems (200; 201).

C. ENTROPY MODELS

The systems optimization techniques just reviewed may be considered to be aspects of control theory, a branch of systems analysis which involves the regulation of the behavior of a system by controlling the structure of its elements and the intensities of flows between them.[8] Another approach to the study of systems is founded on a converse set of principles. Rather than control (i.e. purposive behavior) being exerted over a system so as to bring about a particular state, the notion that the state assumed by a system in any situation tends to be its most probable (maximum likelihood; maximum entropy) configuration is the basis for the family of entropy models.

The various meanings of the term entropy and the utility of each in geography are considered in some detail in chapters 4 and 8. The intent of this section is simply to place within the context of recent developments in systems thinking one aspect of the entropy concept – entropy maximizing models. The notion that many of the regularities that may be observed in the spatial relations of various types of systems result from stochastic processes was introduced into geomorphology by Leopold and Langbein (184), who argued that empirical regularities in the branching of rivers represent maximum entropy, a maximum likelihood distribution of energy. This spatial formulation followed the earlier work of Simon (273) on the non-spatial aspects of systems. Simon postulated that certain size distributions, such as that assumed by the 'rank-size' distribution of urban places, are a manifestation of a set of stochastic relationships (see chapter 8).

In the mid-1960s, Wilson (359) began work that employed the concept of the maximum entropy of a spatial system as a 'useful and practical model building tool.' He developed the entropy formulation into a method of estimating the probabilities of a distribution, given a certain degree of prior information. The result of this work has been a theory for developing various sorts of interaction models in geography. The gravity model, long recognized as empirically useful but lacking in theoretical justification, has been demonstrated by Wilson to be a derivation of the probabilistic entropy model from statistical mechanics; the tenuous foundation of the model in deterministic Newtonian mechanics need no longer serve as the only rationale for the use of the gravity model.

The entropy maximizing method in effect assigns equal probability to any state of a system which is not excluded on the basis of prior information. The most probable state, that which may be achieved in the largest number of ways, is referred to as the maximum entropy state because it corresponds to the position where we are most certain about the microstates of the system since there are the largest possible number of such states in it and we have no grounds for choosing between them (359, p. 6).

As the entropy formulation suggests that regularities that may be observed in the structure of spatial systems are not a function of any organizing principles, it is the deviations from such regularities that are worthy of increased attention since they reveal underlying order and organization. The maximum entropy distribution represents, in a sense, a null hypothesis distribution of spatial behavior.

Entropy modelling involves the maximization of an entropy function subject to certain constraints. For this reason, entropy modelling is directly related to linear programming. This connection is made explicitly by Senior and Wilson (269) for the case of residential location models. They demonstrate that while entropy maximizing residential location models represent sub-optimal systems behavior, this sub-optimality may be more realistic for its specific context than an optimum state.

D. GENERAL SPATIAL SYSTEMS THEORY

A further manifestation of systems thinking in geography, one which is explicitly concerned not simply with systems but with the properties of general systems, derives from the work of Warntz (333; 337; 346; 347). In attempting to develop a conceptual framework for theoretical geography, Warntz expands upon the notion of the coequal duals of 'spatial structure and spatial process' (form and movement), identified by Bunge (55, p. 211) as a way of reconciling Schaefer's (264) laws of morphology with process laws. In examining form and movement, Warntz gives 'special but not exclusive' (333, p. 1) emphasis to the significant geometric and topological properties of surfaces taken generally as a step toward understanding spatial structure and spatial process as general phenomena in neutral terms, 'apart from either the so-called physical or socioeconomic non-spatial properties, for example, attributed to the phenomena' (337, p. 92).

Recognizing the similarity of his approach to the study of space occupying systems and Bertalanffy's (28) attempts at a unified system of knowledge (discussed above in chapter 1), Warntz has identified his style of geographic inquiry as General Spatial Systems Theory (337). The primary methodology involves model building in the mathematical–logical sense; the basic goals involve the attainment of the highest degree of explanation, prediction, and the demonstration of both the unity of the discipline of geography and of our knowledge of spatial structure and spatial process. The philosophy underlying such inquiry is identified by Warntz (333, p. 1) in the following way:

> It is considered to be a valid and important intellectual endeavor to attempt to determine whether or not there exists a certain (small) number of systematic spatial patterns of structure and process among phenomena that may be judged to differ greatly in terms of their non-spatial aspects.

With the tantalizing comment that spatial systems are 'at once a root of philosophical and logical thought and a basis for translating thought into action' (333, p. 1) and some suggestions as to what other types of concepts a General Spatial Systems Theory might usefully employ (337, p. 125), Warntz leaves this notion still undeveloped.[9]

In the judgment of this author, a general spatial systems approach may be one method of achieving a unified framework for organizing the inquiry conducted within the discipline of geography. Such a framework represents a desirable goal for geography since it may have two consequences of some importance. First, it may bring about an end to the fragmentation of the discipline. The major dichotomy within geography, the human and physical branches of the discipline with the many topical and areal sub-specialties within each branch, would be transcended by a philosophy and methodology that is generally applicable in all aspects of the discipline. It is apparent that since the fragmentation of geography has been along the artificial boundaries created by an over emphasis upon the non-spatial properties of phenomena, the unity of the discipline must be reconstructed in a manner that stresses the purely spatial properties of phenomena which may be considered in neutral terms.

A second and even more significant result of the development of a unified geographic framework might be the transcendence of the divisions between disciplines. Geography is an integrating science employing the general theme of spatial relations to break down the walls of the arbitrary compartments into which knowledge has been apportioned. From a spatial viewpoint, at least, the unity of knowledge about reality may be demonstrated. Thus, the ultimate objective of the search for a unified framework is the creation of a general spatial science. Such a general spatial science need not be restricted merely to a theoretical treatment of the subject matter that is presently perceived to be within the boundaries of human and physical geography. A truly general spatial science must expand its perspective beyond traditional geographic concerns and consider a wide variety of spatial phenomena at a variety of terrestrial scales. Similarly, a general spatial science must enlarge the range of its methodologies to include those from other areas of knowledge. The work of the group of geographers responsible for the *Harvard Papers in*

Theoretical Geography, especially Warntz and Woldenberg (347; 365; 370), recent studies by Stevens (288), Eichenbaum and Gale (100), and the contributors to the *Towards a Theoretical Biology* Symposia (330), principally Thom (306), Wolpert (372), and Elsdale (103) may be cited as spatial inquiry involving a wide range of phenomena and methodologies.

Since these consequences are viewed by the author as important ones having implications both for geography and, more generally, for science, the remaining portion of this book will continue the attempt initiated in chapter 1 to impose some degree of formal structure upon the conceptual framework introduced by Warntz.

Notes

[1] Stewart (289; 290; 291; 292), and Stewart and Warntz (293) had previously attempted to apply general laws of physics to geographical systems, but did not explicitly conceptualize these as systems approaches. (This is discussed below.) In the same year as Chorley's paper, Ajo (6) and Leopold and Langbein (184) published research utilizing systems concepts. These studies were more concerned, however, with specific problems rather than with a broad framework for research.

[2] This report was prepared by a committee chaired by Ackerman and composed of Brian Berry, Reid Bryson, Saul Cohen, Edward Taaffe, William Thomas, and Gordon Wolman.

[3] As noted in chapter 1, Langton (177) makes similar criticisms, stating that the general and general theoretic components of systems research have received disproportionate emphasis and the specific properties of individual systems not enough.

[4] See Berry and Pred (27) for a bibliography of some of the earlier works.

[5] A convenient notation for handling dimensions and dimensional analysis is introduced in chapter 5.

[6] Constraints:

$$\sum_{i=1}^{n} X_{ij} = D_j \ ,$$

where D_j represents the demand requirements (column totals); and

$$\sum_{j=1}^{m} X_{ij} = S_i \ ,$$

where S_i represents the supply capacities (row totals).

[7] See Massam (207) for an excellent introduction to the subject of optimizing administrative regions.

[8] See Bennett and Chorley (19) for an excellent discussion of control theory in a geographic context.

[9] Implicit manifestations of a general spatial systems approach can be found in Bunge (55), Anuchin (10), Nordbeck (220), while more explicit treatment is given in Woldenberg (365; 368; 369).

PART TWO

ELEMENTS OF A GENERAL SPATIAL SYSTEMS FRAMEWORK

4

Spatial Structure (I) Morphology: Geometry

In attempting to develop the concept of spatial process and to establish its central position within the general spatial systems approach to geography, we must first consider the two elements of spatial process: spatial structure and movement in space. It is not the goal of the present work to treat structure and movement in a truly comprehensive manner, as this would require, at the bare minimum, a separate and quite lengthy document. Further, such a treatment would be largely redundant as Haggett (140), Haggett and Chorley (141), Bunge (55), March and Steadman (203), and Lowe and Moryadas (191), to cite just a few authors, have considered structure and/or movement at great length. Rather, since one of the aims of a general spatial systems approach is to present a small number of widely applicable concepts, chapters 4, 5 and 6 seek to conceptualize form and movement at a level of abstraction that may be utilized in the description or the analysis of spatial process. The concepts presented represent the most general and basic manner of approaching structure and movement, and the use of these concepts aids in distinguishing between the spatial and the non-spatial (which, for our present purposes, may be regarded as extraneous or of secondary importance) aspects of terrestrial phenomena. The subject of this chapter and the next is spatial structure, the components of which are morphology and dimensionality.

I. Morphology

The word 'morphology' originated with Goethe in his 1817 essay, *Zur Naturwissenschaft uberhaupt, besonders zur Morphologie,* in which the term was proposed for the study of the unity of type of organic form.[1] The morphology of Goethe has been extended to include the wider study of structure, which deals with the forms that are theoretically imaginable. The study of form may be either descriptive or analytical (in the sense not of reduction but of the search for general principles; see chapter 1), the descriptive proceeding from the verbal to the mathematical, and the analytical being concerned with the dynamical relations of form; that is, with the study of the forces that give rise to any particular form (203, p. 270). Chapters 4 and 5 are concerned with ways of conceptualizing and describing structure, while chapters 7 and 8 are concerned with spatial process, the interaction of structures with the movements that give rise to them.

A distinction is generally made between the form of an organism, its shape, and the configuration of some class of phenomena. In actuality, however, there is no distinction; it is purely a matter of the scale at which the inquiry is conducted. What is the shape of some variety of algae but the configuration of numerous individual cells taken in aggregate? The choice of this example is not purely arbitrary, for some of the most insightful examinations of spatial structure have been carried out by biologists in their investigations of the growth of organisms.[2] Similarly, what is the form of an urban area but the collective arrangements of people and their social and economic artifacts?

Within the discipline of geography, morphology may be taken to include the distribution or configuration of phenomena on the surface of the earth and the resulting macro-structures or shapes. As discussed in chapter 2, the substance of geography may be broadly defined so as to include small-scale phenomena as well as global patterns, so that the study of morphology may be seen to include more than the arrangement of market towns in southern Germany. As the reader will recall, the argument was made above that there may exist more individuals engaging in geographic research than there are formally trained geographers; the biologists cited here, for example, qualify as geographers according to this definition. Again, this lends strong support to the argument that there exists an integrating general spatial science that transcends the boundaries of the conventionally recognized disciplines. It should be the task of geography, the only discipline wholly concerned with space and spatial arrangements at a terrestrial scale, to elucidate upon this integrating science. This is the goal of the general spatial systems approach.

Both Sauer (263) and Schaefer (264) used the term morphology in their methodological discussions of the discipline of geography. The former employed the term in a very broad sense in his 'Morphology of Landscape', so as to include the resultant form of a multiplicity of interactions between man and his environment. The latter argued that the laws of geography are strictly morphological (see chapter 1). By this, Schaefer meant that the discipline is concerned with the spatial distributions of phenomena on the earth's surface as they are found, ready made, and not with the processes by which these phenomena came to be located as they are. Schaefer thus distinguished between explanation involving only space (morphology) and that involving only time (process); the latter, he claims, is not within the boundaries of the discipline of geography. This distinction was examined in chapter 1, and was demonstrated to diminish when the concept of spatial process is employed.

Morphology may be seen to include geometrical and topological elements, and while these may be discussed in isolation, independently of movement in space, we must always recognize that spatial structure and movement are inseparable; even 'static' distributions involve movements taking place over long time periods.

A. GEOMETRY

Geometry may be defined as the science or branch of mathematics which investigates the properties or relations of magnitude in space (228, p. 1133). A geometer, then, is one skilled in geometry and, moreover, if the word is taken literally, one who measures the earth. In its earliest form, geometry was concerned with the measurement of physical objects, including portions of the earth's surface (5, p. 12). Thus, the *Arpedonapts,* the land surveyors of ancient Egypt, may have been the first geometers (113, p. 26). Georgescu-Roegen (113, p. 26) identifies geometry as the first theoretical science, consisting as it does of a set of propositions arranged in a logical (and, therefore, economical), as opposed to taxonomic or lexicographic, order. He suggests that the *Arpedonapts* must have come to use, however unawares, the logical algorithms of geometry long before Euclid set them down in his *Elements,* simply because the use of the algorithms saved them the effort of memorizing.

1. The Development of Geometry

Geometry began as the study of physical objects but soon evolved into the study of the space occupied by physical objects rather than the particular objects themselves. Experience with computations, for example the volume of a basket of grain or a block of stone, led to the emergence of a new idea, that the volume was not a property of the grain or of the stone as such, but of the space occupied by the grain or stone. This space, an abstraction from our experience with physical objects, is an aspect of the physical world in which we live, so we refer to it as physical space (5, p. 15).

The physical space occupied by a physical object is part of the physical world. Consequently, geometry as the study of physical space is really a branch of the science of physics. Geometry became a branch of mathematics only when it was transformed into the study of *mathematical space,* a conceptual structure obtained through the idealization of physical space. This transformation took place in intimate connection with the development of the philosophy of Plato (431 BC to 351 BC). As he expressed it in the dialogue called Timaeus, 'That which is apprehended by intelligence and reason is always in the same state; and that which is conceived by opinion with the help of sensation and without reason is always in a process of becoming and perishing and never really is.'[3]

Euclid (*c.* 300 BC), too, dealt with abstract mathematical space, rather than with physical space. His book, *The Elements,* included a systematic presentation of much of the geometric knowledge of his day and has been the basis for the teaching of geometry for over two thousand years (5, p. 63). *The Elements* deals with pure ideas and their relationships with each other and represents the first known attempt to organize formally an entire branch of human knowledge into a deductive system, a collection of propositions in which the aim, to the extent that it is possible, is to define each term that is used and to validate each proposition by proving it by logical deduction from other propositions in the system (5, p. 64).

As in the case of most other disciplines that employ the tools of geometry, geography has been dominated by Euclidean geometry to such a degree that for many centuries it was never questioned as being the one and only language suitable for discussing geographic problems (147, p. 191). Some of the problems raised by location theory, the spatial expression of which is frequently discussed in Euclidean terms, have raised interest in what might be termed social space. Such space often appears non-isotropic when judged by Euclidean standards, and the processes operating in that space often seem to demand a different metric for discussing spatial relationships. In short, geographers have begun to explore the spatial languages developed by geometers as alternatives to that of Euclid in the belief that such languages may provide a more appropriate means for discussing geographical problems (147, p. 191). For example, geographers have begun to explore work like that of Roberts and Suppes (255, p. 175), who point out that our visual perception is not in the straight lines of Euclid, but rather in the constant negative curvature of Lobachevskian geometry; the ability to function in a Euclidean world is probably learned rather than innate.

It was Gauss (1775-1855) who suggested that the uncurved plane surface of Euclid, the flat space on which geometry had hitherto been operating, was only a special case of a more general geometry that could just as well be applied to curved surfaces (80, p. 69). The two primary alternatives to the planar geometry of Euclid are the hyperbolic geometry of Lobachevsky and Bolyai, and the elliptic geometry of Riemann. Both of these geometries demonstrate that different sets of axioms yield different geometries, but geometries which are every bit as consistent as Euclid's (147, p. 199). This simple conclusion dealt a shattering blow to the whole philosophical system of mathematics as set up by Kant, and to much of the framework of rational scientific thought. Where Euclidean geometry had the advantage of being easily and usefully interpreted and applied to a wide range of phenomena, the initial difficulty with the non-Euclidean geometries was to provide them with any kind of interpretation, let alone application, for they possessed properties far removed from direct experience. In short, the non-Euclidean geometries are highly non-intuitive.[4]

The intent of this discussion is not to trace the development of the science of geometry in painstaking detail. Some familiarity with the development of geometry is useful, however, for several cogent reasons. First, as Harvey tells us, the whole practice and philosophy of geography depends on the development of a conceptual framework for handling the distribution of objects and events in space (147, p. 191). At its simplest, this amounts to some coordinate system to give absolute location to objects and events. This may be, for example, a Cartesian system for use on a small portion of the earth's surface, an area that may be considered, for all practical purposes, planar; or, on a larger scale, one might employ latitude and longitude. In more complex instances, this may involve the determination of geodesics on some conceptual surface, perhaps through the use of hyperbolic geometry. Such systems are, in a strict sense, languages, and the geographer must necessarily resort to a language appropriate to his purposes.

Second, the historical development of the substance of geometric inquiry, from physical objects to physical space to mathematical space, is instructive within the context of the theme of the present work, the feasibility and utility of a theoretical approach, an abstract, generalized, and coherent approach, to the discipline of geography.

Third, the methodology of the theoretical geographers in holding constant the non-spatial properties of phenomena exemplifies the abstract and logical reasoning of the early geometers. Just as it was recognized that volume exists independently of grain, or other volume-filling matter, so it should be recognized that a spatial distribution may be considered independently of the substance or specific characteristics of the phenomena so distributed; in terms of the examination of the arguments advanced by Sack (see chapter 2), this means that a right angle and its hypotenuse may exist independently of the living room which they characterize. Physical objects and the physical space occupied by them may be idealized through abstraction so that the 'things which the figures represent', rather than the figures themselves, become the object of our consideration. It is only when this step is taken that the economical and generally applicable operations of mathematics and logic may be applied towards the ends of geographic explanation. Thus, it is suggested, pure ideas and their relationships with each other are suitable and, indeed, necessary (but not sufficient) as the substance of a conceptually unified geography.

Bunge (55, p. 174) has suggested that the development of theoretical geography seems to be recapitulating the history of geometry. Bunge's statement may be interpreted to apply at two different levels. First, at the literal level, Bunge is pointing out the heavy emphasis in theoretical geography upon the logical operations of geometry in describing, classifying, and explaining distributions of phenomena on the earth's surface. Clearly, Bunge has been heavily influenced by the writing of Schaefer (264) on spatial relations and morphology. Also, as with geometry, the course of theoretical geography has run from the more elementary to the more complex forms of analysis. On a second and more metaphorical level, one might point out the similarity between geometry and geography in terms of the respective trajectories of each through a two dimensional phase space, the axes of which are time and a continuum of relative degree of 'vergency': convergence or divergence, unification or fragmentation (Figure 4.1).

Geometry, through the course of its development, has shifted between a fragmented or pluralistic conceptual framework, exemplified by the emphasis on the measurement of individual physical objects and the

A. GEOMETRY

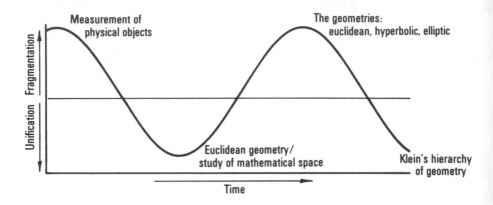

B. GEOGRAPHY

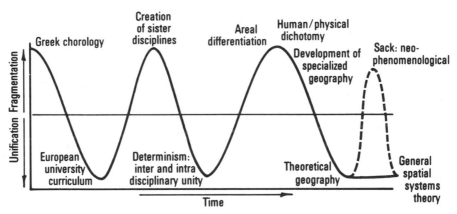

Figure 4.1 *Unification and fragmentation in the substance and structure of geometry and geography*

development of the non-Euclidean geometries, and one of unity, manifest in the concept of mathematical space and Klein's hierarchy. The latter helped to bring about the conceptual unification of geometry by providing a framework which synthesized the various geometries into one consistent and coherent system, based entirely upon the concept of transformation. The hyperbolic and elliptic geometries were demonstrated to be forms of projective geometry in exactly the same sense as is Euclidean geometry.

In the same manner, the development of geography has been a history of swings between pluralism and unity. Figure 4.1 sketches only the barest skeleton of this phenomenon but depicts the general trend. We may identify instances of fragmentation of both the substance and the structure of the discipline: the chorological view of the Greeks and the areal differentiation tradition; the division of the discipline into its 'sister' fields; the human-physical dichotomy; and the development of a host of specialized topical 'geographies'. And we may identify periods in which the emphasis has been upon the unity of the discipline: the curriculum of the early European university; the rise of deterministic thought; and the rise of theoretical geography and general spatial systems theory.

It is for these reasons that a portion of this chapter has been devoted to a brief examination of the development of geometry. This having been accomplished, let us explore more closely the relationship between geometry and geography.

2. Geometry and Geography

Warntz tells us that geometry and geography along with graphics, a related field of considerable relevance to each of the former, had their first grand synthesis in the second century with the work of Ptolemy (339, p. 20). Ptolemy began his *Geography* with the statement that 'Geography is a representation in picture of the whole known world together with the phenomena which are contained therein' (348, p. 16). Having made the distinction between geography and chorography, the detailed description of smaller regions of the earth, and having dealt with questions concerning the shape and size of the earth (he believed it to be spherical but about one-third smaller in circumference than it actually is), Ptolemy devoted several chapters of Book I of *Geography* to the problem of presenting a picture of the known world on a plane surface, that is, to the methods of map projection.[5] Largely on the basis of the rediscovery of the work of eighteenth and nineteenth century scientists such as Gauss, Euler, Halley and Cayley, the modern geographer has begun to recognize that geometry represents much more than just a means of constructing map projections, and that the *geo* in geometry and the *geo* in geography have more than an incidental association. Bunge (55), for example, argues for the necessity and the efficiency of recognizing and operationalizing the inseparability of geometry, the *language* of spatial relations, and geography, the *science* of spatial relations.

There exist numerous examples of instances where geometrical solutions have been applied to problems that are not spatial or where the problems have been abstracted from space and the geometry is employed only by analogy (339, p. 2). In economics, for example, indifference curves are commonly employed to portray a surface of consumer satisfaction; various paths on this surface have meaning with regard to the income effect and the substitution effect. In chemistry, Josiah Willard Gibbs pioneered the use of surfaces and paths to portray relationships in thermodynamics. His geometrical 'phase space' representations enable one to diagram the nature of water, for example, as it undergoes changes from solid to liquid to gas. And, generally, the mathematics of response surfaces, supported by appropriate statistical measures, now appear in many non-spatial disciplines such as psychology (339, p. 2).

Geography differs from disciplines like those cited above in that the *geo* in geometry is taken literally and the language of spatial relations is applied to the examination of the phenomena and processes that exist on the surface of the earth. Reiterating a point raised in chapter 2, the use of an abstract language such as geometry allows the geographer to examine simply and economically the multiplicity of objects and events on the earth's surface, highly diverse in terms of their non-spatial properties, through the utilization of a small number of concepts and techniques. Further, such a language enables the investigator to identify commonalities among the spatial structures of the objects and events that may not be readily evident when their non-spatial aspects are emphasized. Herein lies the key to the unity of the discipline of geography and the impetus for the arguments advanced here that geography may be regarded as a general spatial science that transcends the arbitrary boundaries in the universal set of human knowledge.

3. Geometrical Properties

Bunge (55, p. 212) suggests that spatial structure can be defined most sharply by interpreting 'structure' as geometrical. What type of properties, then, may be identified as characteristically geometrical and, by extension, as components of spatial structure? This section will consider some of the most common geometrical properties, especially those that may be viewed as particularly relevant to the application of geometry to geographical problems.

a. Dimension

Although the concept 'dimension' has several meanings (as will become evident when dimensionality is treated in chapter 5 as an aspect of spatial structure that is separate from geometry), it is related in no small way to a consideration of geometrical properties; the definition of some of the most widely applicable geometric constructs depends upon it. The basic attribute of geometry is length; a line has only the property of length and is, thus, of one dimensional measure (L^1). An area is two dimensional, length taken to the second power (L^2); and a volume has dimensions of length to the third power (L^3). A point, however, has no length and is of zero dimension.[6]

Points, lines, areas, and volumes all have direct physical correspondences in the phenomena examined in geography. For example, depending on the scale employed a coal field or a city might be represented as a point, an area, or even a volume. Similarly, a major highway might be represented as a line or an area. In addition, we may combine some numbers of lines, or lines and points, to form a network (Figure 4.2).

Note that dimension is not really a geometric property in itself but a useful abstraction that may be employed to define some basic geometric constructs. As such, it is distinct from, but related to, the more specific geometrical properties treated below.

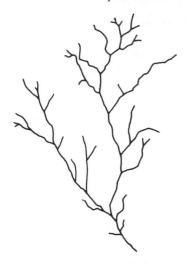

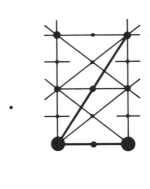

A. Lines: lightning stroke.

B. Points and lines:
road network in an undeveloped country.

Figure 4.2 *Networks*

b. Distance

As noted above, the basic attribute of geometry is length. When considering the spatial distribution of objects and events on the surface of the earth, we generally need to employ a measure of the length or the extent of space lying between phenomena; this length is referred to as distance.

Watson (349) wrote of geography as a 'discipline in distance' and referred to the operational problem of defining a means for measuring distance, a metric. In the absolute space of Kant (see chapter 1), the metric employed must remain isotropic and constant; the only metric available in such a situation is that of Euclidean geometry (147, p. 210). Elliptic geometry, the geometry of the sphere, may be regarded as an extension of Euclidean geometry where distance is concerned. Although the shortest distance between two points on a sphere is the arc of a great circle rather than a 'straight line,' the metric is constant over the physical surface, and therein lies the similarity to the Euclidean metric. Hyperbolic geometry, the geometry of the pseudosphere, may be distinguished from the Euclidean and elliptic forms in that the metric is not constant.

Watson argued that distance can and must be measured by systems that are alternatives to the constant physical metric when concrete geographical, rather than abstract geometrical, problems are considered. For example, in location problems distance might best be approximated by cost rather than by miles; in shopping behavior studies distance might be perceived in terms of travel time. Using such definitions of distance, and taking into consideration mode of transportation, it is possible to demonstrate that Los Angeles is closer to Chicago than is some point that is physically situated between the two cities. It would probably require a smaller expenditure of time and money to get to Chicago from Los Angeles than it would from Crosby, North Dakota, almost 50 percent closer as measured along the arc of a great circle. Similarly, in diffusion and migration studies, one might need to consider distance in terms of amounts of social interaction and intervening opportunity. The latter measure of distance differs from the others cited in that it is an ordinal rather than a cardinal measure; places are ranked as being closer or farther away than other places (55, p. 179).

The realization that the 'best' or most appropriate metric in a particular situation is variable came as the result of attempts to compare theoretical patterns, the central place hierarchy for example, with patterns observed in the real world (147, p. 210). The distance measures employed with the observed patterns were derived using Euclidean concepts of distance, while the theoretical patterns clearly referred to some other measure such as those indicated above. The difficulty of comparing actual with theoretical patterns led to the notion of employing map transformations in order to match patterns measured according to some alternative metric system (147, p. 211). Bunge (55), Tobler (314; 315) and Getis (115) have been the pioneers in this attempt. Operationally, what is involved is making reality conform to the theoretical framework by employing a projection that will transform some aspect of the physical spatial relations. For example, distance (and thus area) might be distorted in such a way that high population densities are expanded and low densities are contracted so that the requirement of uniform distribution of population is met. Thus, a hexagonal market system may be discovered where the Euclidean metric indicates that none exists (Figure 4.3). Although this technique has proven to be extremely useful and revealing, there is the inherent danger that its use may represent an attempt to force nature into a set of inflexible boxes.

A related concern with the variability of the distance metric has arisen in the discipline of psychology in the form of multi-dimensional scaling (MDS). Recently, MDS has found its way into the geographic literature and has proven to be of a high degree of utility for a number of problems considered by geographers.[7] The basic problem in multi-dimensional scaling is how to represent the distances between objects located in some n-dimensional space (147, p. 315), each dimension representing some facet of a complex attribute. The technique is subject to a good deal of interpretation, particularly as the ability to measure distance on a complex multi-dimensional attribute depends entirely upon an ability to specify the geometric characteristics on the n-dimensional space formed by the n dimensions of the attribute (147, p. 334). A generalized form of the Euclidean distance formula is employed even though the objects between which distances are measured are in a 'psychological' space which may be quite different from any known geometrical space. In this formula, the metric may be varied by changing its exponent, the Minkowski constant (120, p. 30). Generally, either a Pythagorean or a 'Manhattan' ("city block") metric is utilized in the determination of distances between stimuli.

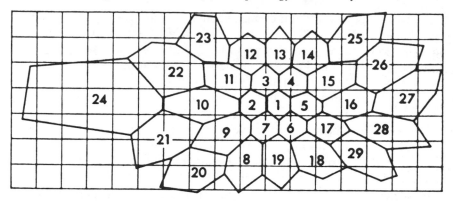

A. An approximation to Christaller's theoretical model in an area of disuniform rural population density.

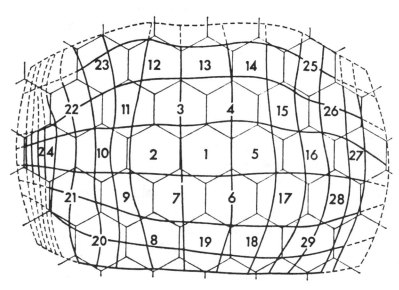

B. A transformed map to produce even population densities and the superimposition of the theoretical Christaller solution.

Figure 4.3 *Comparison between theoretical and actual map patterns by means of map transformations*
Source: Bunge (55, pp. 278–9)

i. Geodesics. The preceding discussion should make it reasonably clear that geographic distance can no longer be equated solely with great circle distance or, at the smaller scale, Euclidean distance. The relevance and utility of alternative definitions of distance can perhaps be best emphasized by reference to the concept of the geodesic path. A geodesic is a minimum path between two points; on a plane it is a straight line and on a sphere it is the arc of a great circle. Man has come to realize, however, that operational space on the earth's surface is not at any one moment to be

87

conceived as representing the single geometric unity of the physical surface of the earth (335, p. 7). The theory of surfaces is equally applicable in many cases to phenomena that are otherwise generally acknowledged to be significantly different in terms of their non-spatial properties. Thus, the modern geographer conceives of surfaces (and distances on them) based not only upon physical features but also upon social, economic, and cultural phenomena. These types of surfaces may be termed conceptual surfaces and may be regarded as overlying the physical surface of the earth. The geometrical and topological characteristics of these surfaces, as transformed, describe aspects of the real world (338, p. 2). Rarely do the paths of least distance as defined in terms of time, cost, energy and so forth on these conceptual surfaces correspond completely to the great circle arcs on earth's physical surface. Yet, these paths may be regarded as geodesics with respect to the surfaces on which they lie. However much they bend and turn with respect to the earth's physical surface as the datum, they do not turn at all but, rather, move straight forward with respect to the surface in whose terms they are defined. 'One man's crooked lines are another man's geodesics' (335, pp. 7-8).

Warntz (338) provides a number of examples of geodesic paths on conceptual surfaces. Figure 4.4 shows some of the family of minimum routes on the time surface integrated about New York City for DC-8 jet aircraft flying to Europe at a constant pressure altitude of 300 millibars on October 17, 1960. Minimum time routes for the return trip would differ, varying with the direction and velocity of winds, and pressure pattern. When an aircraft is flying between two airports separated by a broad expanse with winds of varying direction and velocity, almost never will the great circle route afford the minimum time route. Generally, it is possible to deviate from the great circle route to enhance the speed over the earth's surface by utilizing more favorable winds. This will be attempted, ordinarily, so long as the addition to speed is proportionally greater than addition to distance (338, p. 8). It is only as aircraft become capable of greater and greater speeds that the minimum time and great circle routes begin to converge.

A related problem in the determination of a minimum path arises when the problem of the sonic wave or 'boom' created by an aircraft exceeding the speed of sound is considered. Contrary to popular opinion, the boom does not occur only momentarily at that instant at which the plane achieves the speed of sound but is, rather, a shockwave that travels along behind the plane throughout the duration of its flight as long as the speed of sound is exceeded. A very real problem for a socially conscious aviation administration to consider, then, might be the routing of supersonic aircraft so as to minimize the number of persons 'boomed' enroute. Figure 4.5 indicates a 'minimum man-boom path' from Albuquerque, New Mexico, to Seattle, Washington, on the basis of certain assumptions about type and speed of aircraft and width of its sonic wake.[8]

Additional examples are Eckhardt's (98) least-effort/least-distance

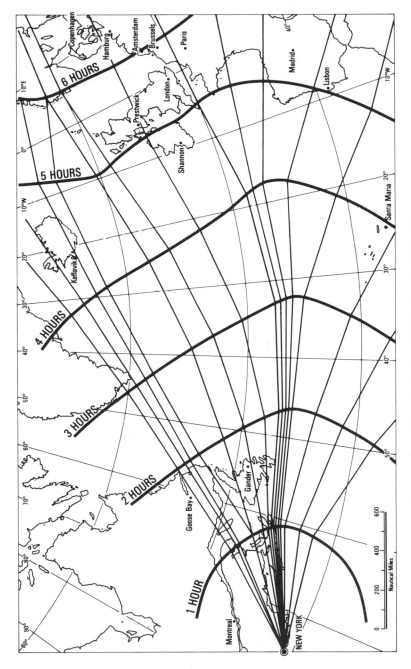

Figure 4.4 *Time surface and routes from New York City, DC-8, October 17, 1960*
Source: Warntz (338, p. 9)

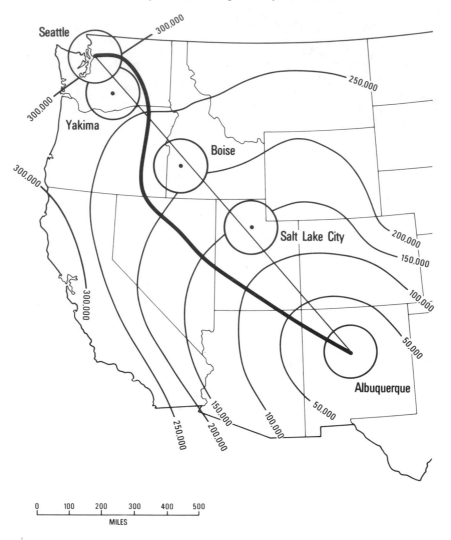

Figure 4.5 *SST minimum man-boom flight path, Albuquerque, New Mexico to Seattle, Washington*
   Source: Warntz (334, ch. 6)

paths followed in the migration of the arctic tern and Bunge's (55) optimal time path for a hiker through a swamp (Figure 4.6). In all of the above illustrations, the geodesic paths run orthogonal to the contours of the conceptual surface; that is, down the gradients. The geodesics could be represented equally well, of course, by straight lines on some appropriately transformed surface. This is illustrated in Figure 4.7.

   ii. Refracted Paths. In the examples above, the geodesics are not straight line or great circle paths over the earth's surface but 'bend' in such a way that they are 'straight' on some conceptual surface overlying the

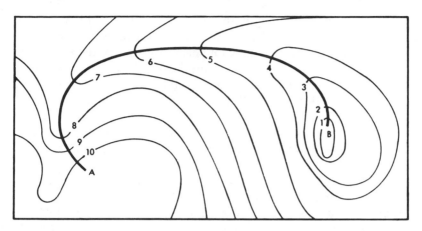

Figure 4.6 *Optimal path for hiker through swamp. Numbers indicate minutes of travel time from B*
  Source: Bunge (55, p. 128)

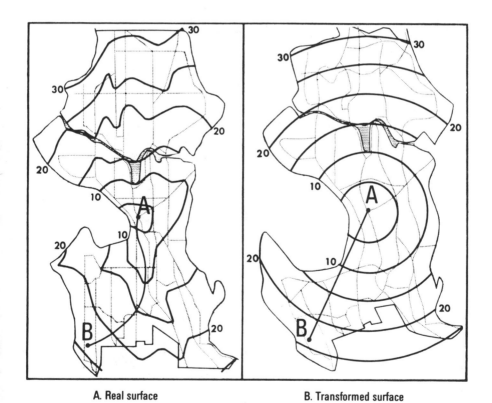

A. Real surface   B. Transformed surface

Figure 4.7 *Real and transformed maps of travel time from central Seattle*
  After Bunge (55, p. 55)

physical surface. These bending paths may be considered within the more general concept of refraction. The law of refraction governs the bending of a light or sound ray when it crosses the boundary from one medium to another as, for example, when leaving air and entering water. Investigation of the refraction phenomenon dates back at least to Ptolemy (*c.* AD 130), but it was Willebrod Snellius, Professor of Mathematics at Leyden, who discovered in 1621 that light falling on a refracting surface at any angle (the angle of incidence) is refracted to a certain angle (the angle of refraction) in such a way that the sine of the angle of incidence maintains a constant ratio to the sine of the angle of refraction (334, chapter 6). This is 'Snell's Law' and the ratio is a number called the index of refraction and may be regarded as a property of a medium itself. In order to establish specific numerical values for a variety of mediums the 'standard' atmosphere is taken as unity. Thus, the index for water is about 1.33 and that of glass about 1.5, and so on. The law of refraction was subsequently refined by Newton and Huygens but remains essentially in its original form (334, chapter 6).

An illustration from the work of Warntz (334; 343) makes explicit the role of a refracted path as a geodesic (Figure 4.8). The general problem is presented in Figure 4.8A; a unit of cargo is to be sent from point A to point B via water and then land and must minimize total transportation cost. Line CD represents a segment of coastline on which some point is to be selected for transshipment. Selection of the least-cost route can be made only when the freight rates, cost per unit per mile, are known for both the

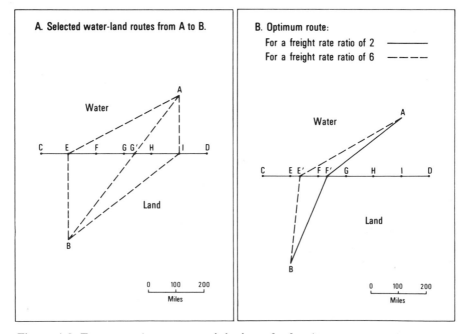

Figure 4.8 *Transportation routes and the law of refraction*
   After Warntz (343, p. 5)

land and the water. The location of the optimum route depends upon the ratio of the land rate to the water rate. Even in the absence of precise knowledge of land and water freight rates, however, it may be demonstrated that the optimum route will not lie outside the quadrilateral AEBI (343, p. 3). If freight rates over land and water are equally costly the least cost route would coincide with the physical geodesic, line AG'B. If the water rate were zero and the land rate some positive amount, the optimum route becomes AEB. If the land rate were zero and the water rate some positive amount, the optimum route would be AIB. In other words, if the refraction index is one, the geodesic is AG'B; if the index is greater than one, as is most likely, the transshipment point will be found between E and G'; and if the ratio is less than one, the transshipment point will lie between G' and I. Figure 4.8B shows the least-cost route from A to B when the cost of traversing a land mile is twice that (solid line) and six times that (broken line) of covering a water mile.

August Losch (190) suggests that this formula for refraction is universally valid in economic and social situations when two unequally favorable zones have to be crossed so as to minimize some objective function. The curved paths of the previous examples may be conceptualized as the result of a sequence of refractions caused by a series of mediums, each having a refractive index that is unequal to that of its neighbors (Figure 4.9). The direction in which a ray of light is bent depends upon the order in which the mediums of relative density occur. For example, a ray going from a less dense to a more dense medium, say from air to water, will be reflected toward the perpendicular; one going from a more dense to a less dense medium, water to air, will be bent away from the perpendicular (257, p. 22).

The examples considered thus far involve the determination of the minimum path between two points, an origin and a destination. A related class of problems concerns the determination of a minimum path system that interconnects a set of nodes. In the early nineteenth century, Jakob Steiner demonstrated that when three soap bubbles meet they do so in a manner that forms 120 degree angles, as this minimizes the length of contact surface between them. Steiner subsequently generalized these results and proved that systems of partitions with three-way corners and, hence, 120 degree angles, use less material than any other systems. Figure 4.10A shows Steiner's original three node problem; B and C demonstrate that this generalization holds for a road network connecting five cities and a sixty-two node military weapons sytem; D illustrates the existence of 120 degree angles in the minimim system of paths connecting 121 points.

A further and somewhat fanciful twist to the determination of geodesic paths is suggested by Warntz (335, pp. 15-16), who asks us to imagine a time when the economically effective demographic energy has increased greatly on the earth's surface, when the existing technological and financial situations have been such that the earth has been nearly smoothed and levelled, when 'at long last in human history the earth has become

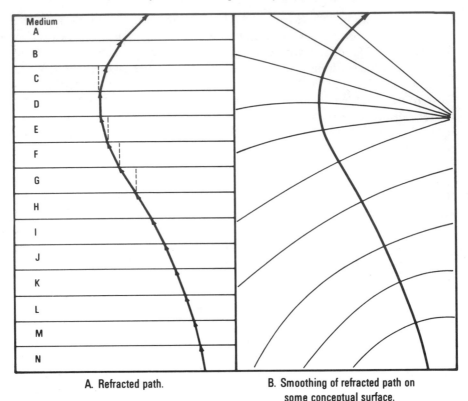

A. Refracted path.

B. Smoothing of refracted path on some conceptual surface.

Mediums N to G are of decreasing density,
Mediums F to D are of increasing density.
Mediums C to A are of decreasing density.

Figure 4.9 *A curved path as a result of a series of individual refractions*

meaningfully spherical.' Under these conditions, virtually all time and cost geodesics will have evolved into simple physical distance geodesics. Further, the actual physical mass of the earth itself might then be regarded as a barrier to travel and the shortest distance between two points might be recognized to be the chord of a great circle arc; that is, a tunnel through the sphere. By dropping into airless, frictionless tunnels in a perfectly spherical earth of uniform density with no Coriolis force a vehicle would begin to fall with maximum acceleration and zero velocity. As it moved to the mid-point of its route, that point where the tunnel is closest to the earth's center, the velocity would increase and acceleration would decrease. Beyond this point, the negative acceleration would cause velocity to decline but the kinetic energy gained during positive acceleration would suffice to permit the vehicle to 'coast up' to the other side where it would reach its destination just as its velocity became zero (335, p. 16). Regardless of the length of the chord tunnel, the one-way trip in any one of them would require 42.2 minutes, a round trip 84.4 minutes. (This latter figure is precisely the one estimated by Newton for a cannon ball fired

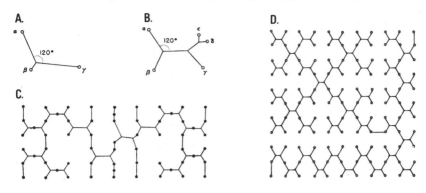

A. Steiner's problem;
B. Five cities;
C. Sixty-two point weapons system;
D. One hundred twenty-one point system.

Figure 4.10 *Minimum systems of paths interconnecting nodes*

parallel to the earth at its surface to orbit the earth, forever falling toward the earth but going just fast enough so as to always miss the earth's surface as it fell.) Every place on the earth, then, would be equidistant from any given point in terms of a time metric, as long as tunnels existed to connect the points.

iii. The Steinhaus Paradox. Finally, we may end our discussion of distance by noting a general problem involved in the use of a physical metric. Steinhaus (284) has pointed out a paradox involving the length of any empirical line. Empirical lines are generally not rectifiable; the tangent to the curve is generally not continuous at all points on the segment and thus there are many points of discontinuity. The paradox is that the more accurately an empirical line is measured the longer it gets. The series of lengths obtained by repeated measures with finer and finer instruments does converge to a finite value but perhaps only at the molecular level (31). In general, the magnitude of the increments of change between successive measures does not vary systematically (223, p. 6). The length of a river or of a coastline may be approximated by summing straight line segments between points on the line. Longer and longer lengths are obtained as the points are chosen closer and closer together. Warntz has summarized the problem by noting that the length of a segment of coastline will vary even at one instantaneous moment depending upon whether it is measured by a person in an aircraft carrier or in a canoe.[9] Perkal (233) illustrates this same idea by the example of a plane circle floating on top of the water and guided by a swimmer such that at every moment at least one point on the circle is in contact with the land (Figure 4.11). Bays narrower than 2ε will be generalized. Obviously, the level of generalization will be a function of the length of the radius. Thus, Perkal introduces the concept of the

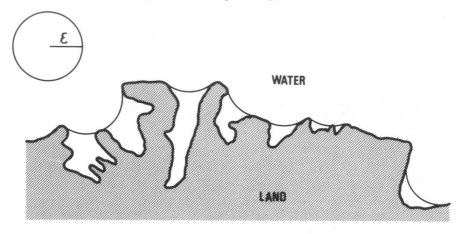

Figure 4.11 *Generalization of coastline measurement*
   After Perkal (233)

ε -length (epsilon length) of a line. Length may be given an unambiguous expression when a given level of generalization, in the one form of an epsilon length, is specified.

The paradox is not to be confused with the fact that all physical quantities are subject to errors of measurement. The problem remains regardless of the level of accuracy achieved; a finer measure of length will always be longer. Nor should this problem be thought of as a theoretical curiosity. No agreement as to the length of an empirical line can be expected unless the scale of measurement is made explicit; this is rarely done (223, p.7). Table 4.1 illustrates the extent of disagreement regarding the length of a mutual boundary between various nations. Naturally, the scale of a map that is used to make a measurement will affect the results obtained. Table 4.2 shows the length of a segment of seacoast between two points on the peninsula of Istrii measured on maps of various scales. Both Steinhaus (291) and Perkal (232) have suggested methods of measurement of length that somewhat reduce the effects of the paradox by including a consideration of scale.[10]

As we have seen, there are many ways of defining and measuring distance. The most useful metric to employ in any situation will vary with the theory and purpose of the investigation. Physical distance, straight line or great circle, is still perhaps the most generally useful metric utilized in geographic analysis. Although the alternative definitions of distance undoubtedly provide more accuracy and a higher degree of explanation, physical distance has been demonstrated to provide a relatively high degree of utility and to serve as a valuable first approximation in most instances. It is for this reason that the argument concerning the role of space in explanation was made in chapter 2. The notion of distance also may be applied, in a non-metric sense, to discrete systems. This is considered in the following chapter in the discussion of topology.

Table 4.1 *Disagreement on length of land frontier between selected nations*

Land frontier between	Kilometers as stated by the	
	former country	latter country
Spain and Portugal	987.0	1214.0
Netherlands and Belgium	380.0	449.5
USSR and Finland	1590.0	1566.0
USSR and Romania	742.0	812.0
USSR and Latvia	269.0	351.0
Estonia and Latvia	356.0	375.0
Yugoslavia and Greece	262.1	236.6

Source: Richardson (253), p. 169.

Table 4.2 *Map scale and generalization of length measurement*

Map no.	Scale	Length in km
1	1:15,000,000	105.0
2	1: 3,700,000	132.0
3	1: 1,500,000	157.6
4	1: 750,000	199.5
5	1: 75,000	233.8

Source: Perkal (233), p. 4.

c. Orientation

Literally, orientation means the act of placing some object so that it faces the east. By extension, it involves placing something in any particular way or ascertaining the position of something, with respect to the cardinal points of the compass or other defined data. In considering orientation, two separate problems are involved. The first is the construction of a system of measurement of orientation. Generally, orientation is defined with reference to some baseline; on the earth's surface the baseline is conventionally taken as a north-south line, but some other line referenced to some other data will serve the same purpose. Orientation is then specified as an azimuth, an angle measured in a clockwise direction from true or magnetic north, or from some other reference point in a different system.

Once a system for measuring orientation has been devised, the second problem needs to be considered; this is the matter of determining the specific direction in which an object is oriented. A conventional way of determining orientation is on the basis of the direction of the longest axis of an object. This is a simple enough procedure with a long thin object but, as Bunge (55, p. 69) points out, this is a highly arbitrary method when some sort of figure is involved since a figure might have several axes almost as long as the longest, and a minor variation in axis length could lead to a major variation in orientation. Bunge suggests two methods of reducing the arbitrariness of the determination of orientation. One involves placing the object in a field of parallel lines and measuring the total length of parallel lines covered by the figure. These lengths may be viewed as a frequency distribution. The field of parallel lines may be rotated several times and the frequency distribution observed. If there is no change in the frequency distribution, the figure is a circle and has no orientation. The more nearly the figure approaches a line, the greater will be the differences in the frequency distributions as the field of parallel lines is rotated. The ratio of these differences could be taken as a measure of the orientability of the figure and the position of the field of parallel lines that yields the least variance could be taken as the direction of orientation (55, p. 69).

A second suggestion by Bunge for reducing the arbitrariness of the determination of orientation is through the use of fixed point theorems. In geometry, a fixed point theorem deals with the points that do not move when a figure is distorted. If a figure has two fixed points, a line can be drawn through them and this non-arbitrary line used, rather than the longest axis, to determine the orientation of the figure. Bunge warns, however, that fixed point theorems do not seem too good a prospect because of the apparent inseparability of orientation and shape. In fact, he notes that it is difficult to conceive of a measure of orientation which is independent of a consideration of shape (55, p. 69). As we shall see next, it is likewise often difficult to produce a measure of shape that is independent of orientation.

d. Shape

i. Open Links. Haggett and Chorley (141, p. 57) make a useful distinction between the problem of describing the shape of lines which represent deviations from a straight path between a given origin and destination (open links) and that in which the origin and the destination are at a common point and the line forms a loop or the boundary of a figure. On the earth's surface, open links generally have their geographic manifestation in stream courses or transportation routes, and several methods of describing their shape are available. Smart and Surkan (277, pp. 965-6) distinguish between two main types of deviations from a straight line path in fluvial systems: wandering paths, characterized by 'all sorts of unsystematic deviations', and meandering paths, defined as curves of considerable symmetry whose dimensions are proportional to the size of the channel.

One of the simplest measures of the shape of a 'wandering' open link is the sinuosity index, the ratio of the observed length (O_L) of the link to the length of the expected path (E_L), where the latter is measured as the straight line distance. Schumm (267, p. 1089) has proposed descriptive categories of sinuosity which range from 'straight' ($O_L/E_L = 1.00$) through the classes 'transitional', 'regular', 'irregular', to 'tortuous' ($O_L/E_L > 2.00$). Figure 4.12 illustrates these categories. Mueller (213) points out that the sinuosity index is somewhat deficient as applied to streams as it fails to distinguish between hydraulic sinuosity, that developed freely by the channel, and topographic sinuosity, that imparted by the geometry of the valley.

The shape of meandering paths poses greater problems of measurement

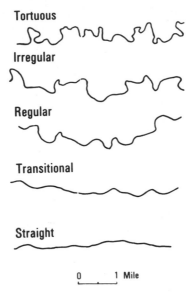

Figure 4.12 *Sinuosity*
Source: Schumm (267)

than does that of the 'wandering' paths. Attention has been focused on the relationships of such meander properties as wavelength (L), amplitude (A), channel width (W), meander belt width (W_b) and mean radius of curvature (R_m). These are illustrated by Figure 4.13. The most consistent relationships have been found between wavelength, on the one hand, and channel width and radius of curvature on the other (141, p. 60). Other approaches to describing the shape of meanders have involved spectral analysis (283) and measures of the probability of a link direction deviating by a certain angle from its previous direction in progressing an incremental distance (176; 263). Clearly, the Steinhaus paradox will have direct applicability to the measurement of both meandering and wandering links.

ii. Closed Links. The problem of defining the shape of closed links or polygons has received more attention than that of the open links. Geographers have always studied shapes but have rarely worked directly with them in a rigorous way. Karlin (166, p. 61) suggests several reasons for this neglect: first, there has been a tendency among geographers to regard a region's shape as the spatial limits of the phenomena under study and thus not intrinsically important; second, until recently there has been a lack of systematic and rigorous methods for defining shape. Those who have investigated ways of defining shape have generally agreed that it is one of the most difficult of the geometric properties to measure. Bunge (55, p. 73) tells us that the primary problem is to invent a measure of shape which: (1) does not include less than shape, as do the measures used by the geomorphologists (slope, elevation, profile and so forth); (2) does not include more than shape, as do the measures used by the mathematicians

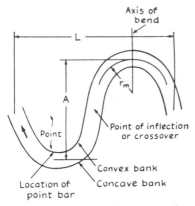

L = Meander length (wave length)
A = Amplitude
r_m = Mean radius of curvature

Figure 4.13 *Parameters in the shape of meanders*
Source: Leopold, Wolman, and Miller (186)

(for example, Fourier series, and Taylor series, and Tchebycheff polynomials, all of which include orientation so that different results are obtained when the surface is rotated); (3) is objective (the common description of shape by devising a classification scheme like 'ox-bow', 'circular', 'shoestring', or 'star-shaped' and assigning shapes to it is highly subjective); and (4) does not do violence to our intuitive notion of what constitutes shape. In summary, the major difficulty in the measurement of shape lies in making sure that the property of shape and no other property is reflected by the measure; the most difficult extraneous property to eliminate is orientation.

Geographers as far back as Strabo (*c.* 60 BC to 20 AD) have recognized the difficulties in dealing with shape. He wrote:

> It is not easy to describe the whole of Italy under any one geometrical figure. Some say that it is a promontory of triangular form . . . . But a triangle, properly so called, is a rectilinear figure, whereas in this instance both the base and the sides are curved . . . . Thus the figure may be said to be rather quadrilateral than trilateral . . . . It is better to confess that you cannot define exactly ungeometrical figures.[11]

Bunge (55, p. 72) notes that the property of shape is present in much implicit geographic theory, running from the geomorphologist's concern for the lay of the land to the urban geographer's classification of the shape of cities. Since Christaller, work on central place theory has been explicit enough to produce testable implications regarding the shape of market areas. De Smith (90, p. 1) views recent geographic interest in bounded shapes as concentrated in three aspects of these finite forms: (1) trip distribution studies, which have been concerned with the extent to which the form of the study region has a constraining effect on possible movement patterns; (2) physical geography, where a variety of studies have utilized measures of shape; and (3) studies of political and economic districting, which have been concerned with the space usage and the effects of inefficient shapes. By an efficient region is meant one in which the ratio of the area to the perimeter is maximized; a circle is the most efficient figure if the region is isolated and not competing for space.

Interest in the measurement of shape has not been confined to geography. The concern with pattern perception in psychology, with the arrangement of gene centromeres in genetics, with petrological, mineralogical, and crystallographic analysis in geology, and with morphometric and morphogenetic approaches to biology all involve the notion of shape and its measurement. In geography and these other sciences, researchers have undertaken a general search for useful, objective, and intuitively satisfying measures, subject to the conditions that the measure be both unique and continuous for a wide variety of shapes. In addition, the measures sought should ideally possess the characteristics of: (1) reversibility, the ability to recover the shape completely from the measure; (2) generality, the ability of the measure to cope with varied and 'poorly behaved' shapes; and (3) flexibility, the ability of the measure to be manipulated by the researcher in a way such that meaningful relationships between pattern and process may be sought (90, pp. 1-2). In these desired attributes of shape, the available measures are often conflicting. Lee and Sallee (182) have stated and proved the following theorem:

'Theorem. There exists no continuous one-to-one function from S, the set of all plane shapes, into R, the set of all real numbers.'

From this theorem, it follows that no single measure of shape will ever be capable of describing complex plane shapes (90, p. 3).

iii. Measures of Polygon Shape. De Smith (90, p. 5) defines shape to be that property of a finite region R, in two-space, which relates to the spatial arrangement of the set of points $\{x_i, y_i\}$ in R, with respect to each other and with respect to the boundary, P, of R. This definition divides shape measures into two classes: those which consider a subset of $\{x_i, y_i\}$, and those which consider the entire set $\{x_i, y_i\}$. De Smith considers the former measures inadequate by his very definition of shape.

The definitions of shape that consider a subset of $\{x_i, y_i\}$ are all simple indices that employ ratios, generally relating the shape of the sampled figure to that of a circle and incorporating a constant term to ensure that the index lies in the range zero to one. Figure 4.14 and Table 4.3 summarize some of the most commonly employed measures. One set of measures involves the ratio of area to the longest axis; the work of Horton, Gibbs, Haggett, Cole, and Schumm (see Table 4.3) falls into this class. In general, these measures overemphasize the importance of the longest axis and, thus, make it possible for the same value to be assigned to very different shapes. A second set of measures, those proposed by Miller, and by Chorley *et al.*, deal with the ratio of perimeter to area. These, too,

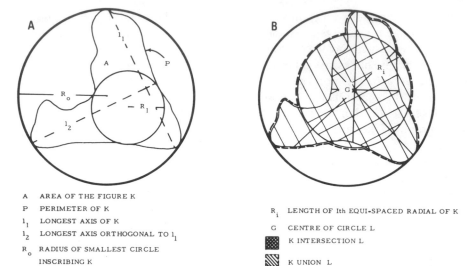

A   AREA OF THE FIGURE K

P   PERIMETER OF K

l_1 LONGEST AXIS OF K

l_2 LONGEST AXIS ORTHOGONAL TO l_1

R_o RADIUS OF SMALLEST CIRCLE
    INSCRIBING K

R_1 RADIUS OF LARGEST CIRCLE
    INSCRIBED BY K

R_i LENGTH OF Ith EQUI-SPACED RADIAL OF K

G   CENTRE OF CIRCLE L

    K INTERSECTION L

    K UNION L

Figure 4.14 *Diagram of shape properties measured by simple indices*
    Source: De Smith (90)

concentrate on a single property of a figure which is indirectly and non-uniquely related to shape (90, p. 6). A third type of measure, that proposed by Lee and Sallee (182) involves the ratio of *K* intersection *L* to *K* union *L* (Figure 4.14B). This is of limited use, as it is difficult to apply and to interpret (90, p. 11). Finally, there is the approach of Bunge (55), which consists of approximating a figure by an equilateral polygon and then summing the distances and squared distances between lagged vertices. The problem with this measure is that if the figure is to be well approximated by an equilateral polygon, the polygon's sides will generally have to be small, and the number of sides and the number of shape sums necessary will become very large. Also, it is possible for equilateral polygons of differing numbers of sides to yield different shape sums for the same figure (90, p. 12).

Among the measures of shape that consider the entire set $\{x_i, y_i\}$ is that proposed by Boyce and Clark (42), who suggest representing a shape by measuring the lengths of equally spaced radii (R_i in Figure 4.14B). They then find the proportional contribution of each radial to the sum of all radial lengths and compare these proportions to those expected in a circle of the same area and with the same number of radials. De Smith views this approach as one which tries to overcome the emphasis of the above measures on extreme properties of the figure. Blair and Bliss (34) argue that a shape may be represented by the set of all distances from its center of gravity to all the infinitesimally small subareas (dA) of the shape (Figure 4.15). Finally, Taylor (305) attempts to characterize shape by the finite frequency distribution of the distances between all pairs of cells of a grid

Table 4.3 Selected shape measures

Date	Author	Formula	Title	Field of study and notes
1932	Horton	A/L_1^2	Form Ratio	Drainage basins
1953	Miller	$A/\pi(P'/2\pi)^2$	Circularity Ratio	Equals $4\pi A/p^2$
1956	Schumm	$2\sqrt{(A/\pi)}/L_1$	Elongation Ratio	Drainage basins
1957	Chorley, Malm, Pogorzelski	P/P_l		P_l is the perimeter of an ideal lemniscate. Drainage basins
1961	Gibbs	$4A/\pi L_1^2$		Drainage basins
1962	Bunge	Shape Sums		US cities
1963	Cole	A/R_1^2	Compactness Index	Mexican villages Civil divisions
1964	Boyce and Clark	$\displaystyle\sum_{i=1}^{n}\left\lvert \dfrac{100 R_i}{\sum_i R_i} - \dfrac{100}{n}\right\rvert$	Radial Line Ratio	Urban form
1965	Haggett	$1.27A/L_1^2$		Brazilian counties
1967	Blair and Biss	$A/\sqrt{(2\pi\int r^2 \mathrm{d}A)}$	Compactness Index	General
1970	Lee and Sallee	$1 - \dfrac{A(K\cap L)}{A(K\cup L)}$	Assignment Function	Nile valley villages
1971	Wilkins and Shaw	$\displaystyle\iint \frac{\sqrt{(x^2 + y^2)}\,\mathrm{d}x\,\mathrm{d}y}{(2/3)(1/\sqrt{\pi})A^{3/2}}$		Urban form
1973	Matwijiw	$AL_2(100/\pi)L_1 R_1^2$		Lancashire political subdivisions

Source: De Smith (90), pp. 8–9.

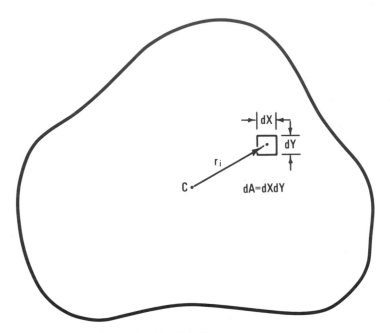

C   Centre of gravity of the figure.

r_i Distance from C to the small element of area dA

Figure 4.15 *Illustration of the Blair and Bliss technique*
  After De Smith (90, p. 17)

superimposed on the figure. He then describes the distributions in terms of their skewness and kurtosis. De Smith (90) extends Taylor's approach to the inferential level by sampling pairs of points within the boundary of the figure and, using the distribution of distances so obtained, producing an estimate of the population distribution which completely characterizes the set.

    Shape, then, may be regarded as a multi-dimensional variable. The number of variables necessary to describe a shape is not fixed or constant, but increases with the complexity of the figure (14, p. 452). Most of the techniques that are available for the measurement of shape are descriptive and cannot be viewed as flexible tools capable of dealing with a variety of fundamental theoretical questions regarding shape. And, again, when dealing with shape we encounter the intrinsic variability manifest in the Steinhaus paradox; this is especially evident in measures such as Bunge's.

    e. Relief
Any physical or conceptual surface may be described in terms of its local relief. Although for most purposes, the measurement of great circle distances for example, the earth may be regarded as a uniform surface, it does have local variability in its relief. Similarly, any conceptual surface may be examined in terms of its degree of relief. Relief is commonly expressed by the concept of gradient, which identifies the change in the

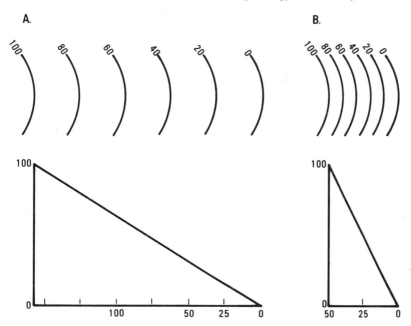

Figure 4.16 *Gradient implied by contour lines*

vertical dimension as the horizontal dimension changes. Thus, we may speak of the gradient on some physical surface, say stream channel gradient or atmospheric pressure gradient, as well as on some conceptual surface such as the income potential or urban population density gradients. On a contour map, where equal interval contours are employed, gradient is depicted by the spacing of the isolines; the closer together the lines, the more rapidly the values change over space, that is the steeper the gradient (Figure 4.16).

f. Pattern

Basically, pattern involves the specification of the spatial relations of discrete phenomena in such a way as to summarize their manner of *distribution* in space. The description of pattern includes some of the problems encountered in the definition of shape. This is not entirely unexpected, as one conventional method of describing pattern is by reference to some shape approximated by the distribution of the phenomena. Fortunately, however, some of the methods available are much more objective than the measures of shape reviewed above.[12] The measures of pattern are quite numerous and range from simple ones, such as the coefficient of localization, to more complex ones, such as spectral analysis. A useful distinction to make among types of patterns is the dimensionality of the phenomena under consideration. Thus, we may differentiate between patterns of points, of lines, and of areas.[13]

Among the measures of point patterns, nearest-neighbor analysis and quadrat sampling are by far the most commonly employed. The former

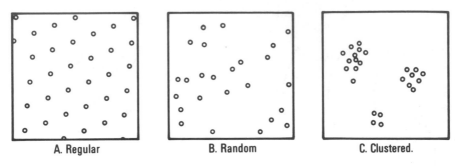

Figure 4.17 *Point patterns*

involves the calculation of the straight line distances separating each point and its nearest neighbor, and the comparison of these distances with those which might be expected if the points were distributed in a random manner within the same area. On the basis of this, point patterns are classified as random, regular, or clustered (Figure 4.17). Nearest-neighbor analysis is recognized as a simple, objective technique for measuring a pattern of points (147, p. 382).[14] The quadrat sampling method is concerned with the probability of finding 0, 1, 2, . . . *n* points in an area of a given size, a quadrat. Here, again, it is possible to calculate the probability under random expectation, a Poisson model, provided that the mean density of points in the study region is known. Then, it is a relatively simple task to measure the deviation from the randomly expected pattern.[15] Both nearest-neighbor and quadrat sampling methods are constructed with reference to some hypothesized mathematical process. An excellent discussion of point pattern is presented by Dacey (87), who raises some fundamental questions concerning conventional methods of analysis.

The description of linear patterns is a much less objective enterprise. One way of classifying linear networks is in simple descriptive terms: dendritic, parallel, trellis, or rectangular (Figure 4.18). Each of these patterns may be broadly related to regional geological structure and may be further modified by local influences (141, pp. 90–1). Bunge and Bordessa (56, pp. 39–42) use a similar typology of road networks in their discussion of the relationship of automobile traffic patterns and injury of children (Figure 4.19). As one proceeds from pattern 1 to pattern 4, through traffic is successively reduced, creating a better and safer environment for children. Another way of treating both fluvial and economic network pattern is in terms of orientation but, as we have seen, orientation is not sufficient to describe shape independently, nor is it sufficient for the description of pattern.

One of the few attempts to depart from a technique of classification according to typed patterns has been that of Dacey (84, pp. 277–9), who proposed a method of discriminating between linear patterns on the basis of their spacing. Figure 4.20B shows a random pattern of lines generated by joining pairs of random points on a Cartesian coordinate system. From

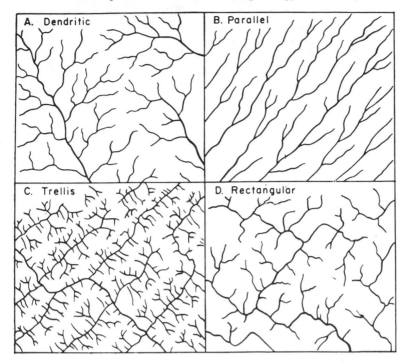

Figure 4.18 *Linear patterns: stream networks*

such a random pattern, lines may diverge from each other until, in the limiting case, they are parallel and orthogonal, a Manhattan grid (Figure 4.20A) or converge to a single point (Figure 4.20C). Average line length and frequency fails to differentiate between the random and non-random patterns, and Dacey proposes the use of intersections on a line traverse, an intersection being formed when the traverse line crosses a line of the pattern (141, p. 102). It is then possible to use a nearest-neighbor type analysis on the intersection points. Tests by Dacey (84) showed clear discrimination between the patterns in Figure 4.20.

Patterns of areas may be of two sorts–implicit and explicit. The former refers to patterns of areas that are imbedded in point or line systems (Figure 4.21), and the latter to some type of 'region' for which clearly delimited boundaries are recognized, a map of political districts, for example. Dacey (86) and Cliff (74) have used contiguity measures to analyze the pattern of areas, while theoretical biologists such as Wolpert (372), Elsdale (103), and Waddington (328) have used a variety of techniques to examine cell pattern and its relation to genetic information.

The study of pattern in all science may be either static or dynamic in approach. The static approach is primarily concerned with describing the geometrical aspects of the pattern. The dynamic approach, on the other hand, seeks to relate pattern to process, to identify the particular process

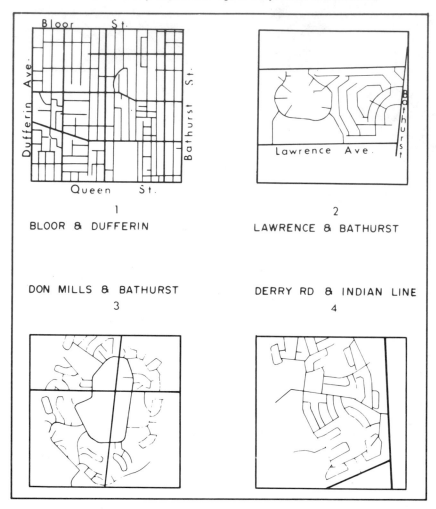

Figure 4.19 *Linear patterns: road networks*
   Source: Bunge and Bordessa (56, p. 41)

which leads to the creation of pattern.[16] As noted, a conventional manner of dealing with pattern is by reference to some hypothesized mathematical process; for example, the randomness of the Poisson distribution. The question that must be seriously considered is whether or not this and similar mathematical processes can be interpreted in terms of some process occurring on the earth's surface. Can the mathematical statement function as an *a priori* model about geographical process? (147, p. 383). This is a difficult question to answer but, intuitively, the answer would seem to be affirmative as many human and physical phenomena behave in a manner consistent with, and homologous to, these mathematical statements. That is not to say that these mathematical statements are themselves processes (see chapter 9). The pattern of the human settlement on an ideal isotropic

A.

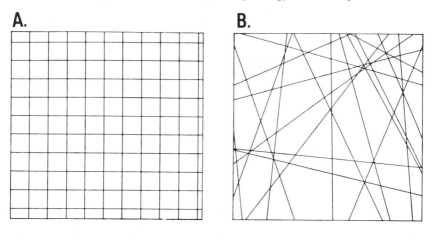

B.

C.

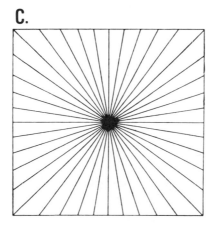

Figure 4.20 *Linear spacing*
  Source: Dacey (84)

surface, for example, may be seen to proceed through a sequence of
random to clustered to regular as it is initially established, seeks the
benefits of agglomeration, and finally undergoes competition for market
area (Figure 4.22). It is of note that the pattern of human settlement under
the conditions of spatial competition is analogous to that of floating
magnets dumped into a tub of water and similarly competing for space (55,
p. 283).[17] Patterns of points, lines, and areas may also imply hierarchies
which may increase the amount of information conveyed to the researcher
about process and organization (Figure 4.23).

  g. Homogeneity
Homogeneity is a rather elusive property to define. A completely
homogeneous pattern or distribution is one in which phenomena are
distributed equally or uniformly with reference to the existing spatial units.
A completely heterogeneous pattern is one in which all phenomena are

A. Areas implicit in transportation network.

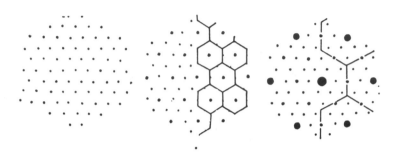

B. Areas implicit in point patterns.

Figure 4.21 *Implicit areal pattern*

grouped in one spatial unit. Between these two extremes there is a continuum of relative degree of homogeneity. The major problem here is in deciding when a distribution changes from a homogeneous one to a heterogeneous one. And what about the notion of randomness? Is a random distribution homogeneous, heterogeneous, or at the mid-point between the two? Obviously, scale plays an important role in the homogeneity concept. The three major regions of the United States, the North, the South, and the West, may be homogeneous with respect to some variable, per capita income for example, but within these regions, say on a county basis, the income may be heterogeneously distributed, that is, locally regionalized. Degree of randomness measured, too, may be demonstrated to depend upon the scale employed.

Methods of describing homogeneity are generally of three types. The first type includes various indices of segregation such as that proposed by Taeuber and Taeuber (303); the second type involves variance analysis to determine whether a unit has a concentration of some variable that is similar to that of its neighbors (380); the third employs the information statistic. The information statistic may be regarded as a measure of organization, as opposed to randomness, in a system and makes use of a basic mathematical notion taken from the second law of thermodynamics in that it regards information as a quantity analogous to entropy. An increase in entropy amounts to a change from a more highly organized state in some physical system to a less organized state. When entropy is maximized, the

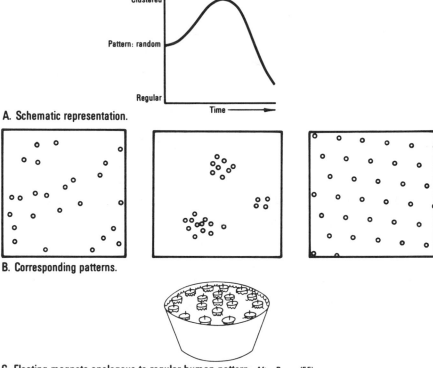

A. Schematic representation.

B. Corresponding patterns.

C. Floating magnets analogous to regular human pattern. After Bunge (55).

Figure 4.22 *Settlement sequence*

system is least organized, more random (Figure 4.24A). As used in information theory, entropy has no relation to its use in the second law of thermodynamics, and an abstract statistical definition of it is introduced. If out of n events, each can occur with the probabilities $p_i = p_1, p_2, \ldots, p_n$, where the sum of all probabilities is equal to 1.0, then the expression

$$H = -\sum_{i=1}^{n} p_i \log_x p_i$$

is called entropy (172, p. 61), where x is some logarithm base. When all events are equally likely, H is at a maximum; when one event is a certainty, H is at a minimum (Figure 4.24A and B). The similarities between this method and those of point pattern analysis reviewed above are quite clear, particularly the emphasis upon the probabilities of various configurations occurring.

Warntz (334) has made a number of important observations concerning homogeneity in his discussion of the so-called 'Sandwich Theorem'. The Sandwich Theorem is an existence theorem relevant to spatial partitioning and derives its name from the example used to popularize the idea: that a sandwich consisting of a slice of bread, spread with butter, and with a slice

111

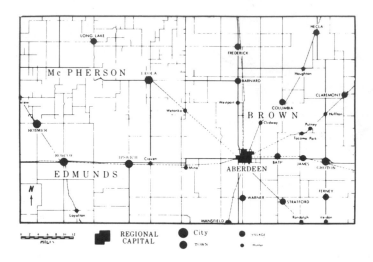

A. Points: City size. Source: Berry (24, p. 31).

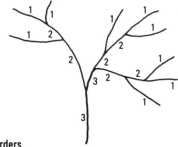

B. Lines: Strahler's stream orders.

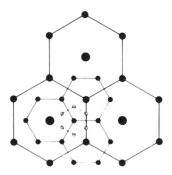

C. Areas: Market regions.

Figure 4.23 *Hierarchies in patterns of points, lines, and areas*

112

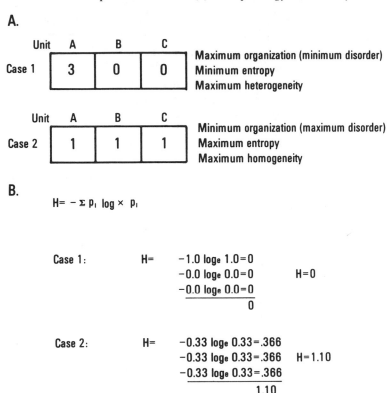

Figure 4.24 *Information statistic*

of meat on it can always be halved (so that each portion simultaneously contains half of the bread, the butter, and the meat) by one plane slice of a knife (334, chapter 2).

The theorem states that,

> Given any three sets in space, each of finite outer Lebesgue measure, there exists a plane which bisects all three sets, in the sense that the part of each set which lies on one side of the plane has the same outer measure as the part of the same set which lies on the other side of the plane.

This was proposed originally by Borsuk and subsequently proved by Ulam. Stone and Tukey recognized it as a particular restricted case and generalized the theorem to consider the nature of partitioning in sets in N dimensions, and partitionings in other ratios in addition to bisections (334, chapter 2).

As an existence theorem, the Sandwich Theorem shows only that a solution exists but does not provide an analytical means of achieving it. However, numerical–graphical procedures may effect sufficiently accurate approximations of the solution for given problems (334, chapter 2). Besides having conceptual relevance to such important problems as redistributing electoral units into areas of equal size and equal populations, the theorem may be used to 'prove' fallacies. Warntz cites the example of the distribution of blacks and whites in the conterminous United States. Using the Sandwich Theorem, he demonstrates that the United States may be divided into 1024 regions, each having about 200,000 persons in it, having an equal percentage of black people. The black population of the United States is thus 'proven' to be homogeneously distributed; racial segregation does not exist. Similarly, income distributions may be manipulated to 'prove' the homogeneity of social welfare. It is possible to defend 'in an irrefutable way mathematically, a certain hypothesis that no poverty regions or racially segregated regions exist on the face of the earth.' Warntz notes that the implications of this are frightening.

h. Location

Location is at once the most elementary and the most complex of the geometric properties. It is the most elementary in that it precedes that which has been defined as the basic attribute of geometry, length. The ability to measure the distance between two points or two objects presupposes the existence of a coordinate system which specifies the absolute or relative location of the phenomena, whether they be points, lines, or areas. The absolute location of some phenomenon must be indicated in such a way that each point on the earth's surface has a unique position with respect to all other points. In general, absolute location may be specified with reference to either a set of spherical coordinates, such as the geographic grid of latitude and longitude with the prime meridian passing through Greenwich, England, or a set of plane coordinates, such as the military grid system.[18] Relative location of some phenomenon may be described with reference to other phenomena, rather than with respect to any fixed global system. Several types of relative coordinate systems are commonly employed, the Cartesian and polar systems for example, and since they are relative an infinite number of specific systems is possible. Many of these relative systems specify location through the use of other of the geometric properties (Figure 4.25). In a dialectical sense, then, location is at once the most primitive and the most sophisticated geometrical property. This reinforces Warntz's (348, p. 254) definition of geography, the most appropriate language of which is geometry, as the science of locations.[19]

4. Conclusion

The interrelationships of the geometrical properties that have been reviewed need to be considered. These properties may be ordered hierarchically on the basis of whether they are primitive or derived concepts. At the top of the hierarchy is location, which is prerequisite to the measurement of distance (length), the basic attribute of geometry. At the second level of the hierarchy are distance (a geometrical but not geographical primitive), orientation, and shape, all of which depend upon some notion of location for their specification. As we have seen, the latter

A. Orientation. Point X is located 5 miles East of Mount Analogue.

B. Shape. Point X is located inside the bend of the river.

C. Relief. Point X is located at the top of the hill.

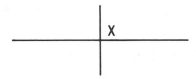

D. Pattern. Point X is located at the crossroads

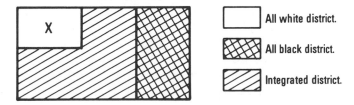

E. Homogeneity. Point X is located in an all white district.

Figure 4.25 *Relative location*

two properties cannot be defined independently of one another. At the third level are relief, a measure derived from distance (both vertical and horizontal), and pattern, which is the arrangement of shapes. Finally, at the fourth level is homogeneity which is defined in terms of pattern and the distance-related concept of scale. As noted, the property of dimension may be applied to the property of length but is not strictly geometric, being of broader applicability (see chapter 5). In combination with distance (length), however, dimension enables us to define points, lines, areas, and volumes which are fundamental concepts to both geography and geometry. More will be said concerning morphological primitives at the conclusion of the section on topology in chapter 5.

Geography is a literal discipline but not in the sense of examining 'real' or existing phenomena. Some of the most profound insights into the spatial relations of concrete objects and events have come through the formulation of models and the analysis of experimentally generated distributions. Rather, geography is literal in the way that it views these real or theoretical distributions. Left and right are physical orientations, not political tendencies; up and down define direction with respect to the earth's center, not emotional states; and close means physical proximity, not emotional attachment.[20] It is this literalness which prompted Warntz and Bunge (345) to title their unpublished manuscript 'Geography – The Innocent Science.'[21]

Geometry may be employed by the geographer as the precise language with which to describe and to analyze the literal properties of actual or hypothesized distributions of phenomena. These literal properties are conventionally portrayed graphically on a map, although they may sometimes not all be represented simultaneously, the case of shape and areal extent on a small-scale map being an example. Even the alternative distance metrics may be portrayed literally through appropriate map transformations. The map is the outward and visible manifestation of this literalness. Thus, we return to the theme of *geography, geometry,* and *graphics,* and may reinforce our appreciation of the interrelatedness of this threesome.

We have seen that geometry began as the study of physical objects but eventually evolved into the study of mathematical space, a conceptual structure arrived at through the idealization of the physical space associated with physical objects. Geography may be viewed as following a similar course, commencing with the study of real world places and phenomena and finally developing, through the use of geometry, into the study of the conceptual structures underlying the distributions, the spatial relations of these phenomena. This is a certain sign of the increased maturity of the discipline.

Notes

1. This reference is cited in March and Steadman (203, p. 38).
2. For example, see Wolpert (372), Elsdale (103), and Thompson (309), to cite just a few.

[3] Quoted in Adler (5, p. 40).

[4] A discussion of the non-Euclidean geometries is contained in Adler (5, pp. 195–260).

[5] For excerpts from Book I of Ptolemy's *Geography*, see chapter 6 in Warntz and Wolff (348).

[6] A fascinating treatment of dimensions may be found in Edwin Abbott's *Flatland* (1), a geometrical–satirical masterpiece which gives us glimpses of life in Pointland, Lineland, Flatland, and Spaceland.

[7] See, for example, Tobler (316; 318; 319) and Golledge and Rushton (120).

[8] See Warntz (334), chapter vi, for a discussion of the assumptions made and the technique employed.

[9] William Warntz, in conversation, November 1974, London, Ontario.

[10] For an introduction to the problem of measuring the length of empirical curves, see Nystuen (223).

[11] Quoted in Bunge (55, p. 72) from C. Van Paassen (325, p. 6).

[12] The distinction between pattern and shape may lie in the distinction between discreteness and continuity. That is, the description of shape seems to be a problem involving a phenomenon having a spatially continuous manifestation, while the description of pattern involves spatially discrete phenomena. This distinction may provide some insight into the relative ease and objectivity with which patterns are described by geographers.

[13] A recent introduction to the study of pattern which follows this approach is that of Getis and Boots (118).

[14] For examples of nearest-neighbor analysis, see King (170), Dacey (83), or Clark and Evans (72).

[15] For examples of quadrat sampling, see Harvey (148), Dacey (85), and Getis (117).

[16] Here, 'process' is used in the conventional sense, as distinct from the definition of spatial process made in chapter 1.

[17] For further discussion of pattern, particularly among naturally occurring phenomena, see Stevens (288) and Thompson (309).

[18] Note that in order to specify *absolute* location, it is necessary to employ coordinate systems which are themselves *relative*. This suggests that absolute location is absolute only in the less rigorous sense of 'by convention'.

[19] Of course, locations cannot be fully understood without reference to movement. As demonstrated in chapter 1, the location of an object may represent a transitory point in the movement of that object through time-space. And, as Bunge notes, 'everything on the earth's surface is where it is having moved there' (in conversation, November 1976, London, Ontario).

[20] In the phenomenological approach to geography, the attempt is to delve beneath the implicit presuppositions of objective analysis and to seek the primitives of understanding. In this approach, the meaning of places to people is sought, and 'close' means an emotional attachment rather than proximity. See Entrikin (104) for a review of the phenomenological concept in geography.

[21] William Bunge, in conversation, November 1976, London, Ontario.

5

Spatial Structure (II) Morphology: Topology; Dimensionality

I. Morphology

B. TOPOLOGY

1. Introduction

We have seen that the term morphology encompasses the general form or the general spatial characteristics of the distribution of any set of phenomena. Further, we have examined the role of geometry as the language which may be employed in the discussion of the *literal* properties of any such distributions; geometry is a literal language with the basic attribute of length. In geography, as well as in science generally, it is often desirable to consider a problem at a higher level of abstraction so as to gain further insight into the more general relationships that exist among phenomena. When such a problem concerns the morphology of a set of 'real' or hypothetical objects and events distributed in some physical or conceptual space there is, similarly, an appropriate language that we may employ. This is topology.

A topological approach to spatial structure involves the reduction of the literal distribution to its most basic and elemental form, its relational, non-metrical aspects. Naturally, when moving from a literal to an abstract representation of spatial structure much information is lost; for example, the distance between two places, the orientation of a transportation route, or the shape of a region. Corresponding to this loss of information, however, is an increase in scope and flexibility. As argued in chapter 1, an approach characterized by a higher level of abstraction is more generally useful in many instances. There are several cogent aspects to this. First, abstraction enables the investigator to increase the number of cases that he may subject to analysis. For example, topological analysis can be used in many situations where geometrical analysis is not possible because no metric is specified. Second, by concentrating on the more elemental relationships one may remove the 'background noise' and discover the commonalities that exist among a wide number of structures. One topological form may be homologous to an infinite number of geometrical forms (Figure 5.1). Third, abstraction enables the attainment of economy and simplicity, which are the goals of science (see chapter 1).

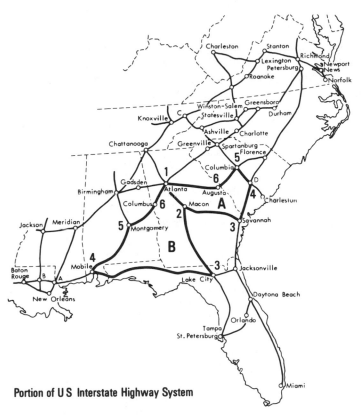

Portion of U S Interstate Highway System

Node Circuit	1	2	3	4	5	6
A	Atlanta	Macon	Savannah	Durham	Columbia	Augusta
B	Atlanta	Macon	Lake City	Mobile	Montgomery	Columbus

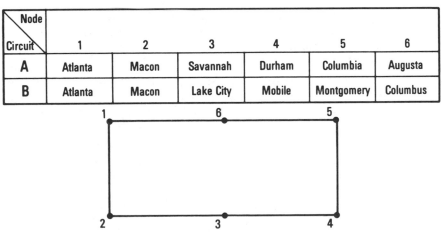

Figure 5.1 *Representation of two geometric forms by one topological structure*

119

2. Topology Defined

The study of topology is considered to have begun with Leonhard Euler in the middle of the eighteenth century, although there is evidence that Descartes had a knowledge of topological relationships during the mid-seventeenth century (106, p. 346). There is some question as to the relationship between topology and geometry. Although generally recognized to have begun as a branch of geometry, topology is considered by some mathematicians as the most general form of geometry (5, p. 371) and by others as having undergone such generalization and having become involved with so many other branches of mathematics that it is more properly considered, along with geometry, algebra, and analysis, as a fundamental and independent division of mathematics (106, p. 345). This question of classification may be considered of relevance in a formal sense solely within mathematics as there is no disagreement as to the scope or substance of topology.

Topology may be regarded as a form of 'qualitative geometry', concerned with the preservation of certain non-metrical relationships such as proximity, separation, order, enclosure, continuity, and contiguity or connectedness. The distinction between these relations and the quantitative properties of geometry (which were enumerated in chapter 4) will be illustrated below. The basic attribute of topology may be regarded, by definition, as continuity and, by extension, the related property of contiguity or connectedness. If we define topology on the basis of some set of transformations, as we did with the various 'geometries' in the last chapter, we find that topology is the study of geometric figures that remain invariant under all unique transformations that are reversible and continuous (5, p. 371; 106, p. 345). Reversibly continuous transformations exist in any space in which the concept of continuity can be defined. Continuity can, in fact, be defined in a very general class of spaces called topological spaces, which include projective spaces and also many other spaces that are not projective (5, p. 374). In this sense, topology is the most general, and therefore the most useful, form of geometry; it is often referred to as 'rubber sheet' geometry because of its emphasis on continuity (Figure 5.2).

a. Topological Space

The term 'topological space' employed above needs further explication. A set of objects called 'points' is called a topological space if it contains a special class of subsets that has the following three properties:

(1) the entire space and the empty set are both members of this special class of sets;
(2) the set obtained by combining the memberships of any number of sets that belong to this special class is also in the special class;
(3) for any two sets that belong to this special class, the set consisting of the points they have in common is also in the special class.

The sets that belong to this special class of subsets in a topological space are called the open sets of the space, or the neighborhoods of the space.

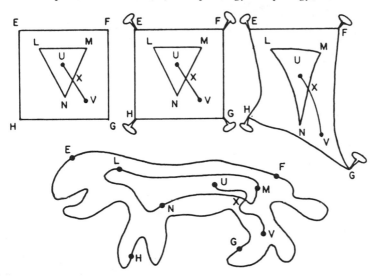

Figure 5.2 *'Rubber-sheet geometry'*
  Source: Cole and King (80, p. 88)

Points that belong to the same open set are said to be neighbors, and neighbors are said to be near each other. Note that this general concept of nearness is defined without invoking any concept of distance between points (5, p. 372).

Under this definition, any collection of objects can be converted into a topological space, usually in more than one way. Adler (5, pp. 372–4) uses the example of the set of four symbols $\{\times, 2, +, *\}$. Its elements are chosen arbitrarily, so that they have no relationship to one another beyond the fact that they happen to have been thrown together in the same set. The set is a loose aggregation, then, and has no structure. The set acquires a structure, however, and the elements become related to each other as neighbors, as soon as we single out certain subsets that will be called neighborhoods or open sets.

As an example, we might specify that these four sets should constitute the class of open sets: $\{\times, 2, +, *\}$, $\{\times, 2\}$, $\{+, *\}$, and $\{\}$ (the empty set). This class satisfies the three requirements listed above:

(1) the whole space and the empty set are members of the class;
(2) the set formed by combining the memberships of any number of sets in the class is also in the class. By combining the memberships of the sets $\{\times, 2\}$ and $\{+, *\}$ we obtain the set $\{\times, 2, +, *\}$, which is in the class;
(3) the set consisting of the common membership of any two sets in the class is also in the class. The set consisting of the common membership of $\{\times, 2\}$ and $\{+, *\}$ is the empty set $\{\}$, which is in the class.

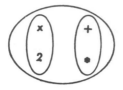

Figure 5.3 *Neighborhoods in topological space*
After Adler (5)

Since the requirements are met, this class of 'open sets' defines a topological structure for the set {×, 2, +, *} and converts it into a topological space. The manner in which the topological structure relates the elements of the space to each other as members of neighborhoods is indicated in Figure 5.3, where each open set that is not empty is represented by a loop enclosing its members.

It is possible to give the same set {×, 2, +, *} another, different topological structure by choosing other subsets to employ as open sets. Thus, many different topological structures may be defined by specifying from among the same set of elements the 'open sets'. Each different specification of 'open sets' is different as a space because in each of the topological structures the elements relate to each other differently as members of interlocking neighborhoods (5, p. 374).[1]

b. Topological Equivalence

Two configurations are said to be topologically equivalent if each can be transformed into the other by a topological transformation; that is, by a transformation involving stretching, shrinking, or bending, but without tearing or welding. For example, a cube is topologically equivalent to a sphere because it can be converted into a sphere by reversible continuous deformation, and every reversible continuous deformation is a topological transformation. A coffee cup is topologically equivalent to a doughnut because each, if made of modelling clay, can be obtained from the other by a reversible continuous deformation (5, p. 375). A sphere is essentially different from a plane because it is closed and finite rather than open and infinite. A sphere can be distorted without tearing into a cube, into a shape like a potato, and so on. The property of such a distortion is, again, that the transformation involved be unique and continuous over all points. Harvey (147, p. 203) uses the following example to illustrate a unique and continuous transformation. Suppose that we map a series of locations on a spherical surface onto a flat plane surface. This, of course, is the classic problem of map projection. The relationships between points on the flat paper may represent spherical relationships locally (that is, the positions of three cities such as London, Birmingham, and Bristol may not be distorted), but on a typical Mercator projection Japan looks to be at the other end of the world from Seattle. At some point in the transformation of points from a spherical surface to a plane surface the neighborhood relationship between points has to be disrupted. The transformation cannot be unique and continuous.

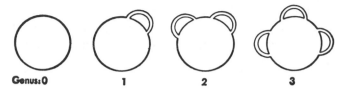

Figure 5.4 *The genus of orientable closed surfaces*
After Adler (5)

If two closed surfaces are topologically equivalent, they must be either both orientable or both non-orientable. Orientability is related to, but distinct from, the concept of orientation and is a dichotomous property; a figure on a surface is either orientable or non-orientable.[2] Similarly, if two closed surfaces are topologically equivalent, they must both have the same genus. The genus of a surface is the largest number of non-intersecting simple closed curves that can be drawn on the surface without separating it into two or more parts. A curve is called simple if it does not intersect itself; thus, an oval is simple but a figure eight is not. The genus of a simple polyhedron or of a sphere is zero, but the genus of a torus, a doughnut shape, is one. Conversely, it can be shown that if two closed surfaces are either both orientable or non-orientable, and if they have the same genus, then they are topologically equivalent. Consequently, the topological properties of a closed surface are determined once we specify its genus and whether or not it is orientable (5, p. 376). Adler demonstrates a simple way of modelling all closed surfaces of finite genus. If p handles are attached to a sphere the resulting surface is an orientable surface of genus p (Figure 5.4). A torus is topologically equivalent to a sphere with one handle, so its genus is one. The surface of a common pretzel shape is topologically equivalent to a sphere with three handles, so its genus is three.

c. Breadth of Application

As one might expect, the abstraction and generality which characterize topological analysis make possible a wide range of applications in very diverse fields. The field of mathematics, of course, has accounted for the major portion of topological applications. An attempt to review the role of topology in mathematics will not be undertaken beyond the brief introduction already provided. The direct application to geographic analysis of some of the most important and most basic of the topological relations discovered by Euler, Cayley, and Clerk Maxwell will, however, be examined in the next section.

In mechanical and, especially, electrical engineering topology is widely employed, as the absolute spatial ordering of the components of some device is generally irrelevant; the connectivities between parts is often the primary consideration. In the social sciences, Piaget (236, p. 205) has suggested that young children form topological concepts before they form geometrical ones. At the age of three, a child can readily distinguish between open and closed figures; if asked to copy a square or a triangle he draws a closed circle. Lewin (189) has long advocated the application of topology to psychology without, however, using the formal mathematical apparatus of topology. In sociology, too, particularly in communications theory, the utility of topology is rapidly being recognized.

Perhaps the most extensive application and development of topological theory outside of the realm of mathematics has been in the field of biology. D'Arcy Thompson's work of 1917, *On Growth and Form* (309), is one of the first applications of topological principles outside of mathematics. Topological relations are implicitly contained in Thompson's observation that skulls and other skeletal parts of different animals can be transformed into each other by simple continuous unique transformations (Figure 5.5). These observations amount, in essence, to the statement that the different shapes of skulls and bones are all derivations of common forms. René Thom, a topologist working on biological problems, notes that there appears to be a striking analogy between the fundamental problem of theoretical biology, the origin and evolution of biological structures or, simply, morphogenesis, and the main problem considered by the mathematical theory of topology, the reconstruction of a global form, a topological space, out of all of its local properties (308, p. 89).

The biophysicist Rashevsky (246) makes explicit the need for topological theory in the attempt to unravel the complex problems studied in theoretical biology. Up to this point, mathematical biology has emphasized almost exclusively the metric aspect of biological phenomena. This is not unexpected since, generally, no meaningful prediction can be made without quantitative expression. When the phenomena that characterize biological integration are examined, however, not quantities but certain rather complex relations are observed. It is Rashevsky's contention that topology can be applied to the examination of not only structural but also functional relations (246, p. 125). All life performs movements that will bring it into contact with certain things such as food, air, and light, and away from other types of things. Rashevsky formulates a method of representing similar biological relations in terms of a topological space; this leads to his principle of 'biotopological mapping' (246, p. 128).

The principle of biotopological mapping states that there exists one, or very few, primordial organisms which may be characterized by their graphs (a graphic representation of relations in topological space; see Figure 5.1 for an example of a graph). The graph of *all* other organisms may be obtained from this primordial graph or graphs by a transformation which contains one or more parameters; different organisms correspond to different values of those parameters. The problem is, therefore, to establish within the principle of biotopological mapping proper hypotheses as to the structure of the primordial graph and as to the transformation laws. In this sense, Rashevsky is trying to do at an abstract level that which Thompson sought to demonstrate at the physical level, that the entire membership of a set may be derived from one of its most elemental members. Rashevsky argues, further, that in biology purely relational or

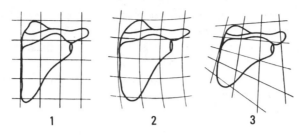

A. Human scapulae. 1. Caucasian; 2. Negro; 3. North American Indian.

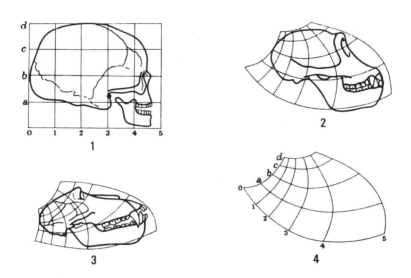

B. Human and animal skulls. 1. Human; 2. Chimpanzee; 3. Baboon. 4. Coordinates of Chimpanzee's skull as a projection of the cartesian coordinates of the human skull.

Figure 5.5 *D'Arcy Thompson transformations*
Source: Thompson (309, pp. 318–19)

topological properties are at least as characteristic of life as the metric properties. Thus, it is natural to consider the possibility that the phenomena of life may be not the *results* of physical phenomena, but that both are the expressions of the geometric properties of the space-time universe, the latter being the expression of metric properties, the former of topological properties (246, p. 126). This approach nicely illustrates the degree of generality sought by the reduction of some set of phenomena to its most basic properties.

Finally, the most recent utilization of topology in science is in the realm of catastrophe theory. This field is the invention of the topologist Rene Thom and is presented in his book *Structural Stability and Morphogenesis* (307). Briefly, catastrophe theory is a mathematical method for dealing with discontinuous and divergent phenomena. This is regarded as a major breakthrough, since the preeminent method of building models to describe and explain events has been, for 300 years, the differential calculus. Differential equations, however, have an inherent limitation; they can describe only those phenomena in which change is smooth and continuous. Relatively few phenomena are that orderly and well behaved; on the contrary, the world is full of sudden transformations and unpredictable divergences which call for functions that are not differentiable (377, p. 65). Catastrophe theory derives its name from the fact that it can be applied with particular effectiveness in those situations where gradually changing forces or motivations lead to abrupt changes in behavior. Many events in physics and biology and the social sciences may be identified as mathematical catastrophes. Zeeman (377) illustrates the following physical and social examples of a mathemetical catastrophe: the transition between a liquid and a gaseous state; the buckling of an elastic beam; sudden changes in the stock market; and abrupt changes in emotional state, for example from anger to fear or from self-pity to anger.

Catastrophe theory is derived from topology. The underlying forces in nature may be described by smooth surfaces of equilibrium; it is when equilibrium breaks down that catastrophes occur. The problem for catastrophe theory is, therefore, to describe the shapes of all possible equilibrium surfaces. Thom has solved this problem in terms of a few archetypal forms which he calls elementary catastrophes. For processes controlled by no more than four factors, Thom has shown that there are just seven elementary catastrophes (377, p. 65). Figure 5.6 illustrates the two simplest catastrophes, the fold and the cusp; the representation of the remaining catastrophes requires four or more dimensions.

The applications of catastrophe theory in geography have been few. This may appear as somewhat surprising for a discipline which has been traditionally interested in both spatial structure and the succession of forms. Perhaps the explanation lies in the problems of relating conceptual and physical space; i.e. mapping the former onto the latter.[3] One would expect, however, that catastrophe theory may have direct relevance for the modelling of any events having a manifestation in real space which exhibit discontinuity in their spatial–temporal structure.

The opening of a new shopping center or the establishment of a growth pole, for example, may represent abrupt perturbations of the spatial equilibrium of a marketing system; the pattern of consumer shopping trips may shift suddenly. Further, any movement down very steep gradients on a physical or conceptual surface may be regarded as a catastrophe. The pattern of migration from rural to metropolitan areas is one example that readily comes to mind. This migration is orthogonal to the isolines on a population or income potential surface – up the steep gradient. For purposes of modelling, potential may be taken as a negative quantity, as it is in physical science, so that the migration is an abrupt movement down the gradient. Similarly, sudden shifts in the position of earth material due

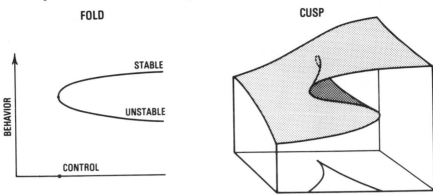

Figure 5.6 *Fold and cusp catastrophes*

to erosion or weathering may be regarded as catastrophes. One of the few explicitly geographical examples comes from the work of Amson (8) who suggests that under certain conditions the density of population in a growing city will show 'jumps' in response to continuous changes in rents and wealth.

Wagstaff (331) employs catastrophe theory to model the settlement of the southern Mani region of Greece. In this region the 'settlement development process' was not continuous, exhibiting a major discontinuity between the late second century AD and the early seventeenth century AD. Change in settlement development is modelled by an 'undesirability function' which, in graphic form, corresponds to an elementary cusp catastrophe. Wagstaff notes that statistical models generally represent settlement change by Poisson and negative binomial distributions, which assume continuous change, generally expansion, in the settlement pattern. The same is true of Medvedkov's entropy model, as operationalized by Semple and Golledge. All three models are adequate for describing continuous settlement pattern development but cannot incorporate discontinuity. For this reason, catastrophe theory appears to be a more appropriate model to employ in the case of the southern Mani region.

Although the manifestations of these events may be modeled using topology, thereby making possible a purely qualitative representation, the underlying factors are still in the realm of spatial process. Catastrophe theory, thus, does not furnish the explanation which scientists seek. Rather, it provides a language for discussing the complex interrelations which need to be understood before explanation can be achieved.

3. Topology and Geography
a. Euler's Formula

A general topological relationship among points, lines, and areas was first established by the Swiss mathematician Leonhard Euler in his 1736 paper on graph theory, in which he introduced the now classic Konigsberg bridge problem. The layout of the city of Konigsberg in Prussia (now Kaliningrad,

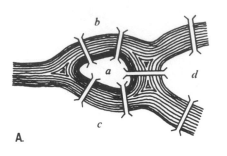
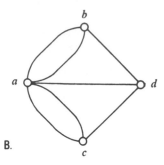

A.   B.

Figure 5.7 *The Konigsberg bridge problem*

USSR) was such that seven bridges joined the four disconnected sections of the city (Figure 5.7A). The problem involved the question of whether it was possible to take a walk around the city starting from any point and to arrive back at the same place having crossed each of the bridges only once. Since the bridges are, for the purposes of the problem, the most important feature, Euler devised a simplified diagram in which the four parts of the city were represented by points and the bridges by appropriate lines joining the points (Figure 5.7B). This simple arrangement of points and lines is the 'graph' of graph theory. This particular graph is planar as no lines intersect except at vertices.

The value of a graph like this one lies in the capacity it has for portraying clearly the essential structure of a set of relationships. In the Konigsberg problem we know that we must arrive in each part of the city (*a, b, c,* and *d*) the same number of times as we leave it, otherwise we would never get back to the same starting point. Since we may cross each bridge (traverse each edge of the graph) once, this means that for there to be a possible route fulfilling the requirements, the number of edges joined to or incident with each vertex must be even. This is true for none of the four vertices *a, b, c,* or *d,* and so the problem has no solution, as Euler demonstrated (202, p. 243). A related problem, also a classic in graph theory, is that of attempting to connect each of three wells to each of three houses in such a way that none of the pipes cross. Again, there is no solution. However the pipes are drawn, there are always at least two that cross (Figure 5.8).

Euler first formulated his graph theory in terms of the vertices, edges, and faces of a simple polyhedron. When the polyhedron is converted by a topological transformation into a figure on a two dimensional surface, the network of edges on the polyhedron becomes a network of lines which divides the surface into many pieces, each corresponding to a face of the polyhedron. Similarly, each vertex of the polyhedron corresponds to a point on the two dimensional surface. Thus, the points, lines, and areas represented by a graph are conventionally referred to as its vertices, edges, and faces. Euler demonstrated that the general relationship $F = E - V + 2$ (the number of faces equals the number of edges minus the number of vertices plus two) holds for any closed surface.

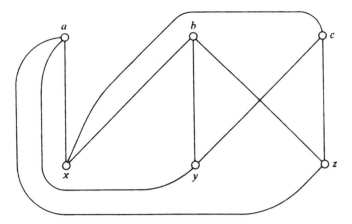

Figure 5.8 *The problem of the houses and wells*

The applications of graph theory in geography are too well known to require any review here. Garrison (112) and Kansky (164), as well as many other researchers, have devised numerous measures for use in the description and analysis of graphs. The majority of these measures are concerned with the centrality of vertices and the degree of connectivity of a network. Distance, too, is a property of concern to the graph theorist, but distance expressed as a topological rather than metric property; that is, the distance between two vertices is expressed as a degree of connectivity, the number of edges that intervene between the two points. The study of the characteristics of transportation networks has been the most frequent use of graph theory in geography. Figure 5.9 shows a transportation network and four equivalent graphs. One of the most powerful and flexible aspects of graph theory is the isomorphism between a graph and a binary matrix indicating connectivities between vertices (Figure 5.10). Such a matrix may be manipulated by matrix algebra to yield further measures. Finally, Bunge has used graphs to illustrate, in a qualitative rather than quantitative manner, some of the realities of human existence (Figure 5.11).

Since the early application of graph theory in geography, represented by the largely structural approaches of Garrison and Kansky, there has been considerable theoretical development of this method of analysis. Much of this recent work has dealt with the flows or 'processes' associated with structural characteristics. Tinkler (311) provides a useful review of both the early and the more recent graph theoretic approaches in geography, noting that contemporary work has taken three specific directions. First, there is nodal analysis in which interaction flows between nodes are investigated (73; 175; 17); second, the analysis of both specified and optimal routes through a graph – the problem of traversibility (121; 150); third, coverings and colorings of graphs (300; 313). Scott (268) provides a review of the last two directions.

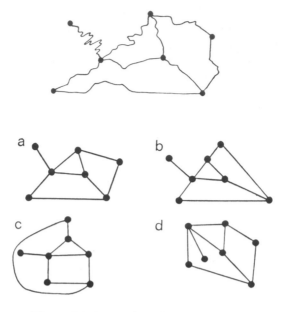

Figure 5.9 *Network and equivalent graphs*
Source: Haggett and Chorley (141, p. 5)

Some of the most novel geographical applications of Euler's work are to be found in Warntz's studies of both social–economic and physical phenomena (340; 341; 343; 346; 347). Following Cayley (62) and Clerk Maxwell (208), Warntz demonstrates that any surface, physical or conceptual, may be reduced to its singular points (its pits, peaks, passes, and pales), its ridge lines (running from peak to peak) and its course lines (running from pit to pit), and the territories composed of portions of hills (areas defined by course lines) or dales (areas defined by ridge lines). Noting the formal isomorphy between these points, lines, and territories and Euler's vertices, edges, and faces, he proceeds to discuss the general topological relationships on both open and closed surfaces (for example, the number of peaks plus the number of pits minus the number of passes plus the number of pales always equals two on a closed surface) and to demonstrate the utility of these relationships in the analysis of the structure of, and the flows on, any continuously differentiable surface.

Warntz and Waters (346) extend this analysis by identifying the critical elements (faces, vertices, and edges) of an atmospheric pressure surface, reducing these to networks, codifying the networks as matrices, and performing appropriate matrix operations on the networks. These matrix operations not only describe the connectivities of the system but also help to suggest solutions to flow problems within it. Parenthetically, it is research of this nature that is demonstrating the unity of the discipline of geography and laying the foundation for the general spatial systems approach.

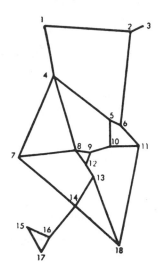

A. Graph.

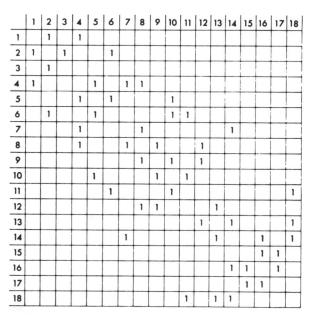

	1	2	3	4	5	6	7	8	9	10	11	12	13	14	15	16	17	18
1		1		1														
2	1		1			1												
3		1																
4	1				1		1	1										
5				1		1				1								
6		1			1					1	1							
7				1				1						1				
8				1			1		1			1						
9								1		1		1						
10					1				1		1							
11						1				1								1
12								1	1				1					
13												1		1				1
14							1						1			1		1
15																1	1	
16														1	1		1	
17															1	1		
18											1		1	1				

B. Binary connectivity matrix.

Figure 5.10 *Equivalence of matrix and graph*
  Source: Bunge (55, p. 53)

A. Land power only.

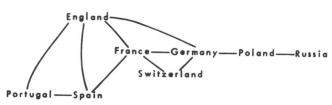

B. Land and sea power.

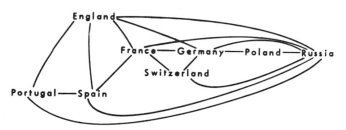

C. Land, sea, and missile power.

Figure 5.11 *Graph of European connections*
   Source: Bunge (55, pp. 108–9)

b. The Four Color Map Problem

One problem of traditional interest to geographers at both the practical and theoretical levels has been that of the contiguity of regions. Arthur Cayley gave this problem a formal mathematical basis with his 1879 paper, 'On the Colouring of Maps' (61). Although a line drawn on a map distinguishes two regions or areal units from one another, both regions cannot be colored the same because it would imply that they are part of the same region; different colors would improve the distinguishing characteristics of the line. The so-called Four Color Map Problem is this: it has been demonstrated mathematically that on a surface of genus zero such as the earth or a plane, five colors are sufficient in order to color a map in such a way that no two regions that share a common boundary (a line) will be of the same color. It has been found from experience, however, that only four colors are necessary to color a map in this manner. This has led not only to attempts to rework the mathematical theorem but also to attempts to construct a map requiring five colors.

As in the case of Euler's Konigsberg bridge problem, the four-color problem is one involving topology; the connectivities between regions are the critical relationships and this may be represented in network form by a graph. Gardner (111, p. 126) notes that a map that requires five colors can be discovered by looking for a graph that has the following properties:

(1) it is connected (all in one piece),
(2) it is planar (edges do not intersect, except at vertices),
(3) it has no bridge (an edge such that if it were removed, the graph would decompose into two disconnected pieces),
(4) it is trivalent (three edges meet at every vertex), and
(5) it is not three colorable (the edges cannot be colored with three colors, one to an edge, so that all three colors meet at every vertex).

The existence of such a graph has just recently been disproved as a consequence of the proof of the Four Color Map Problem constructed by Appel and Haken (12). This proof is suspect, however, in the eyes of some mathematicians in that it is not a conventional 'pencil and paper' short proof, but one which cannot be performed without the aid of a computer. Twelve hundred hours of computer time were necessary to develop the proof. Incidentally, it can be demonstrated that seven colors are necessary and sufficient to color the regions of a torus (genus one) so that no two contiguous ones share the same color.

Tinkler (313) provides an example of the manner in which map coloring problems may be used in the analysis of spatial questions. Rural periodic markets and the market areas surrounding them are conceptualized as a planar graph. Two rural markets are adjacent when they share a common boundary. In order to avoid direct competition between markets, the graph should be colored so that no two adjacent markets have the same color – hold market on the same day. The four-day market week typical of West Africa is interpreted as being directly related to the four-color conjecture. 'Is it too much to suggest that the widely found four-day market week is the natural response to the topology of actual market systems?' (313, p.3).

c. Topology of the Earth's Surface

The earth's surface may, at a first approximation, be regarded as spherical in geometrical terms, and topologically, as a single continuous closed surface. The sphericity of the earth has long been recognized and assigned a role in geography but the topological nature of the surface has not received as much attention. Warntz (334, p. 12) notes that in virtually every case we have posited a flat open plane as our reference surface. The topology of the earth has generally been considered such that any complete simple boundary curve circuit on the open surface separates the area into an included one and an excluded one. On any closed surface that is the topological equivalent of a sphere, a complete simple boundary curve circuit again divides the surface area into two parts, neither of which, however, is included or excluded.

In order to illustrate this, Warntz asks us to regard any very small circle on the surface of a sphere as bounding a certain included area. For example, think of a circular prison wall with an initially small radius. But the arbitrariness of this notion of 'included–excluded' on a topological surface equivalent to that of a sphere is shown if we allow the radius of the wall to increase gradually. At first the length of the prison wall is approximated by the expression $2\pi r$ and the prisoner knows that he is imprisoned. Eventually, however, the expanding radius will have brought the entire surface of the earth, minus one point, behind the prison wall. Only one point, the antipodal point of the original circle's center, remains 'free'. The circumference of the prison wall will have gone from near zero to near zero through a maximum of approximately 25,000 miles. Warntz concludes his example by noting that the topologies of the open plane surface and that of the closed spherical surface are significantly different; on the earth's surface some barrier such as the 'iron curtain' may restrict interaction between two groups of people but it must be recognized that, simultaneously, both groups are behind it or neither is.

Bunge, too, manifests a concern with the recognition of the 'true' topological properties of the surface of the earth. He notes that the 'natural' genus of the earth, zero, is being continually tampered with as tunnels are drilled through the earth. Bridges, too, change the genus of the earth, similar to the addition of handles to a sphere, as mentioned previously. The actual number of tunnels and bridges on the earth's surface has never been estimated; thus, the earth's genus is undetermined (55, p. 110).

Related to Bunge's observation is the plot of a well-known science fiction story, 'A Subway Named Mobius' by A.J. Deutsch (91). A Möbius (or Moebius) strip, named after its discoverer, Augustus Ferdinand Möbius, a German mathematician, may be constructed by taking a long thin strip of paper, bending it in a circle so that the two short edges overlap, and giving one end a single twist before joining them (Figure 5.12). The Möbius strip possesses unusual topological properties; specifically, it has only one side and one edge and behaves in a very counterintuitive manner when it is sliced lengthwise. The story involves the Boston subway system being expanded in such a way as to acquire a very complex set of connectivities. It finally gets so complex as to become a 'spatio-hyperspatial network' suggesting 'whole families of multiple-valued networks, each with an infinite number of infinite discontinuities.' The result of this is that trains begin to disappear into the 'non-spatial' part of the network. The title, of course, is a parody of Tennessee Williams' 'A Streetcar Named Desire.'

4. Geometry, Topology, and Geography: Conclusion

The distinction between geometry and topology may be generally characterized as one of literalness versus abstraction. As we have seen,

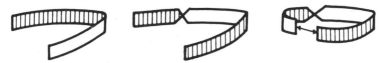

Figure 5.12 *Construction of a Möbius strip*

each of these modes has its advantages; a geometric approach affords a higher level of specificity and information while topology is more flexible and of greater generality. Within geography, the geometric–topological distinction most directly involves the primary tool of the discipline, the map. A geographic map may portray either geometric or topological properties. A 'geometric' map is one that portrays the earth's surface, or some portion of it, literally, preserving (although often not simultaneously) properties such as distance, orientation, and shape. This is the conventional map of the discipline, as it has been much more widely employed by geographers than the 'topological' map, a more highly simplified and non-literal model of the world (Figure 5.13). Topological properties such as proximity, continuity, or connectedness may be inferred from a geometric map, of course, but the procedure is non-reversible.

One form that the topological map commonly assumes is the graph. In general, it is possible to represent any kind of route map as a graph. The actual way in which the graph is drawn is of no significance; an edge represents the existence of a connection between two points but indicates nothing about direction, length, or the shape of the network (see Figures 5.1, 5.9 and 5.13). The edges of a graph may not even represent a physical link (a road) but, rather, may represent an air connection.

A second form of topological map is the more general logic diagram which Gardner (110) defines as a 'two dimensional geometric figure with spatial relations that are isomorphic with the structure of a logical statement.' To this he adds,

> The spatial relations are usually of a topological character, which is not surprising in view of the fact that logic relations are primitive relations underlying all deductive reasoning and topological properties are, in a sense, the most fundamental properties of spatial structures. Logic diagrams stand in the same relation to logical algebras as the graphs of curves stand in relation to their algebraic formulas; they are simply other ways of symbolizing the same basic structure.[4]

If we consider, with Warntz (338), the sets of places on the earth's surface (which is the 'set of all sets', the universe of discourse of the geographer) as *necessarily* having spatial properties and *capable* of having non-spatial properties (climatic, cultural, economic, political and so forth), the relationships between these sets can be shown graphically by the use of Venn-like diagrams. Those sets containing elements, the non-spatial properties of which are alone specified, may be portrayed using the conventional Venn diagram, in which the relative connectivities, rather

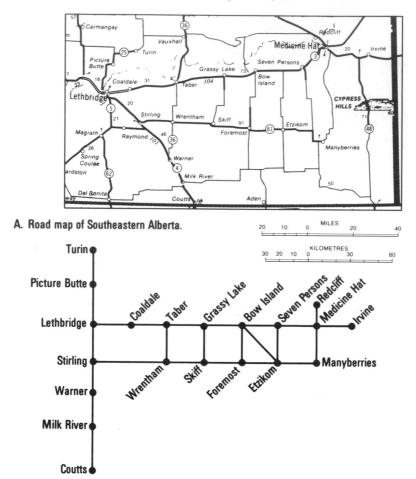

A. Road map of Southeastern Alberta.

B. Topological map of roads between selected communities.

Figure 5.13 *Geometrical and topological maps*

than the absolute positions of elements, are of import. Sets containing elements that have some demonstrable spatial property may be portrayed on an 'extended Venn diagram', the geographical map, in which the literal spatial relationships between elements are crucial. Both graphical forms are maps in the general mathematical sense.

Warntz (338, pp. 14-25) illustrates the relationships of topological and geometric maps in the following way. The universe of discourse of the Venn diagram is conventionally portrayed as a rectangle (Figure 5.14A), but this is a mere convention and any other bounded shaped could be used (5.14B). If L represents the set of all elements (places) on the earth's surface that are regarded as dry land, and if L' represents the complement of L, those places that are not regarded as dry land, this relationship may be graphically depicted in Figure 5.15A. No spatial locations have been

136

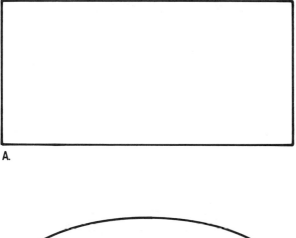

A.

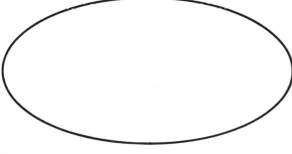

B.

Figure 5.14 *The universe of discourse*

specified. Here, only the connectivities are of import; there is a subset of the earth's surface which is dry land and a subset which is not dry land. These two subsets are contiguous but do not overlap, and they exhaust the universe of discourse. The location of these non-spatial properties may be depicted as in Figure 5.15B. The latter is a conventional geographic map, or in Warntz's terminology, an extended Venn diagram. The simple closed curve representing the universe of discourse has had a geographic grid overlaid and becomes a standard Mollweide projection.

Figure 5.16 illustrates several topological maps of the sets L and L', and H and H', where H is the set of elements having the non-spatial property such that the average temperature of the coldest month is above zero degrees Celsius, and H' is its complement. Using conventional set notation, we may define such regions as L union H (the aggregate of all elements which are members of set L or set H, or both) and L intersection H (the elements which are members of both set L and set H). Figure 5.16D shows the geometric map of this intersection. Warntz further notes that Boolean algebra, the mathematical language of sets, may be usefully employed to perform logical operations on the elements portrayed in the topological map.

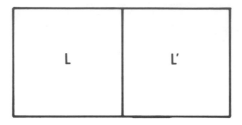

A. Topological map.

B. Geometric map.

Figure 5.15 *Topological and geometric maps L and L'*

A longer and more systematic demonstration of the relationship between the general mathematical concept of the mapping of sets, including both topological and geometric mappings, has elsewhere been completed; the role of Boolean algebra and of the additional graphical tool of the logic tree are also made more explicit.[5] This illustration involves the system of climatic classification proposed by Köppen and modified by Geiger. Under this system, any point of the earth's surface may be placed in an appropriate climate region defined on the basis of temperature, precipitation, and rate of evapo-transpiration. Figures 5.17A and B show the first and final topological maps of climate regions. These Venn diagrams correspond, of course, to the extended Venn diagrams or geometric maps showing: (a) the distinction between type E climates and those climates that are not type E; and (b) the entire earth divided into the twenty-seven possible climate types possible under the Köppen–Geiger system.

In the conclusion to chapter 4 the elementary geometrical properties of distance, orientation, and shape were identified. We may now add the

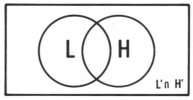

A. Existence of properties L and H.

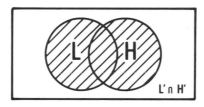

B. L ∪ H.

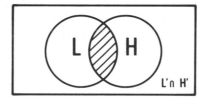

C. L ∩ H.

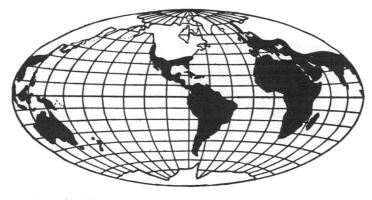

D. Geometrical map of L ∩ H.

Figure 5.16 *Topological and geometric maps of L and H*

basic topological property of connectivity to this list; the resulting set of concepts may be regarded as morphological primitives. As in the case of the elementary geometrical properties, the notion of connectivity is, dialectically, both a function of location and a prerequisite to its specification. Any discussion of morphology, then, must be undertaken with reference to these primitives.

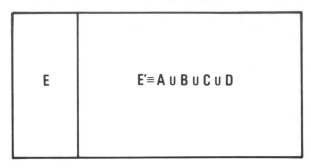

A. Type E and non-E types distinguished.

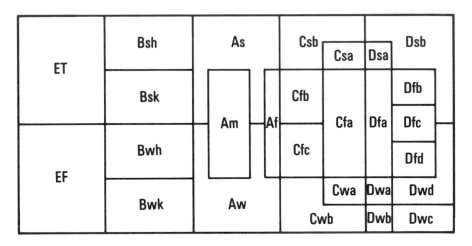

B. Representation of the twenty-seven climate types.

Figure 5.17 *Venn diagrams of Köppen–Geiger climatic regions*

II. Dimensionality

A second component of spatial structure, in addition to morphology, is rather more abstract and involves the dimensionality of the phenomena being investigated. There are two distinct, yet related, aspects of dimensionality. The first of these consists of the identification of the property of dimension of some phenomenon. The second aspect is dimensional analysis, a technique in which the property of dimension is employed in the derivation of theory and the validation of formulae. Each of these two aspects will be considered in turn, and their relevance to geographic inquiry examined.

A. DIMENSION

In 1832 Gauss identified three fundamental physical units or 'substantial' variables: length, mass, and time. These are extensive variables and appear to be elementary scientific notions which can neither be derived from one another nor resolved into anything more fundamental (158, p. 12). All other units are derived units (and intensive) and may be defined only by reference to one or more of the three fundamental units. Derived units depend not only on multiples or sub-multiples of the fundamental units, but also on the powers and the combinations of powers of length, mass, and time. For example, the unit of area is [unit length^2] and unit volume is [unit length^3]; area is said to have a *dimension* of 2 in length [L^2], and volume a dimension of 3 in length [L^3]. (The square brackets signify 'dimensions of'.) The derived unit, velocity, is the ratio of the number representing the distance covered by a moving point to another number representing the time occupied in traversing that distance; that is, a number of length units divided by a number of time units (158, p. 14). Expressed symbolically, this is [LT^{-1}]. Thus, the term dimension has come to be conventionally employed as an abbreviation of the phrase, 'exponent of dimension'. Table 5.1 illustrates the dimensions of some common derived physical quantities.

Clearly, the fundamental physical units mass, length, and time are not sufficient to deal with the substance of geographic inquiry in a comprehensive manner. Stewart and Warntz (293, p. 116), noting that geography is frequently concerned with value and with population size, suggest that these be recognized as primary social units, along with the three primary physical quantities. A truly comprehensive science of the earth, including both human and physical aspects of existence on this planet, would, then, be based upon these five fundamental quantities. This is not to say that this set, particularly the social units, is completely exhaustive. There may be other primitives that have not yet been recognized. Information is a likely candidate for admission to this set. Table 5.2 illustrates the dimensions of some derived variables commonly employed in geography.

The term dimension is also employed in a second fashion, in the sense of a ruler along which different characteristics are measured. Thus, both Stewart and Warntz (293, p. 116) and Haynes (151, p. 52) refer to mass, length, time, value, and population size as 'primary dimensions', and Stewart (290, p. 245) speaks more generally of the search for the 'dimensions of society'. The distinction is primarily a semantic one; the same set of relationships is indicated by both forms of usage.

1. Dimensionless Numbers

Thus far we have established that it is possible to identify primary and secondary (fundamental and derived) quantities, and that the latter are defined in terms of their dimensions in the former. It is also possible to identify pure or dimensionless numbers within many physical and social–economic systems. These dimensionless or natural variables derive their meaning from the physical systems that they describe without reference to any arbitrary physical standards (162, p. 3). Further, such dimensionless numbers serve as criteria for the prediction of system behavior at various

Table 5.1 *Dimensions of derived physical quantities*

Physical quantity	Exponents of dimensions			Formula		
	L	*M*	*T*			
Volume density	−3	1	0	L^{-3}	M	
Length per unit volume	−2	0	0	L^{-2}		
Area density	−2	1	0	L^{-2}	M	
Curvature	−1	0	0	L^{-1}		
Linear density	−1	1	0	L^{-1}	M	
Angle	0	0	0			
Mass	0	1	0		M	
Length	1	0	0	L		
Mass × length	1	1	0	L	M	
Area	2	0	0	L^2		
Moment of inertia	2	1	0	L^2	M	
Volume	3	0	0	L^3		
Speed of density change	−3	0	−1	L^{-3}	M	T^{-1}
Velocity per unit volume	−2	0	−1	L^{-2}		T^{-1}
Momentum per unit volume	−2	1	−1	L^{-2}	M	T^{-1}
Velocity per unit area	−1	0	−1	L^{-1}		T^{-1}
Viscosity	−1	1	−1	L^{-1}	M	T^{-1}
Frequency	0	0	−1			T^{-1}
Mass per second	0	1	−1		M	T^{-1}
Velocity	1	0	−1	L		T^{-1}
Momentum	1	1	−1	L	M	T^{-1}
Kinematic viscosity	2	0	−1	L^2		T^{-1}
Action	2	1	−1	L^2	M	T^{-1}
Volume per second	3	0	−1	L^3		T^{-1}
Acceleration of density change	−3	1	−2	L^{-3}	M	T^{-2}
Acceleration per unit volume	−2	0	−2	L^{-2}		T^{-2}
Force per unit volume	−2	1	−2	L^{-2}	M	T^{-2}
Acceleration per unit area	−1	0	−2	L^{-1}		T^{-2}
Pressure	−1	1	−2	L^{-1}	M	T^{-2}
Angular acceleration	0	0	−2			T^{-2}
Surface tension	0	1	−2		M	T^{-2}
Acceleration	1	0	−2	L		T^{-2}
Force	1	1	−2	L	M	T^{-2}
Temperature	2	0	−2	L^2		T^{-2}
Energy, torque	2	1	−2	L^2	M	T^{-2}
Rate of change of volume per second	3	0	−2	L^3		T^{-2}
Power	2	1	−3	L^2	M	T^{-3}

M = mass; L = length; T = time

Source: Huntley (158), pp. 149–50.

Table 5.2 *Dimensions of derived social quantities*

Social quantity	Exponent of dimension					Formula	
	M	*L*	*T*	*D*	*N*		
Population density	0	−2	0	0	1	NL^{-2}	
Land price	0	−2	0	1	0	DL^{-2}	
Average income	0	0	−1	1	−1	DN^{-1}	T^{-1}
Yield per acre	1	−2	−1	0	0	ML^{-2}	T^{-1}
Man hours	0	0	1	0	1	NT	
Resources per person	1	0	0	0	−1	MN^{-1}	
Transport rate	−1	−1	0	1	0	DL^{-1}	M^{-1}
Dyadic interaction	0	0	0	0	2	N^2	
Population potential	0	−1	0	0	1	NL^{-1}	
Income potential	0	−1	0	1	0	DL^{-1}	
M = mass; L = length; T = time; D = dollars; N = population							

Source: After Haynes (151).

scales (360, p. 11), and are regarded by Fein (107) as indicative of the formulation of a fundamental theory.

The simplest sort of natural variables are ratios. Angle is defined as arc/radius, $[LL^{-1}] = [L^0] = 1$. Similarly, pi, the ratio of the length of the circumference of a circle to its diameter, is dimensionless $[LL^{-1}]$. The Mach number is the ratio of the speed of a moving object to the speed of sound in the medium through which it travels, $[LT^{-1}L^{-1}T] = [LT^{-1}]/[LT^{-1}] = [L^0T^0] = 1$. The first of the more complex dimensionless numbers was identified by Reynolds (252) in 1883 as a means of describing flowing fluids. This is the Reynolds number and represents the ratio of inertial forces to viscous forces. When the numerical value of this natural variable exceeds 2000 for river channels, the inertial forces overcome the viscous forces and turbulence occurs (363, p. 11). The Reynolds number is given by

$$\frac{pVD}{m}$$

where p is the mass density of the fluid $[ML^{-3}]$; V is flow velocity $[LT^{-1}]$; D is a characteristic length of the system $[L]$; and m is the viscosity $[ML^{-1}T^{-1}]$. The dimensional formula, then is:[6]

$$[ML^{-3}]\,[LT^{-1}]\,[L]\,[M^{-1}LT] = [M^0L^0T^0] = 1.0$$

In geography, Strahler (297) has deduced four dimensionless numbers which completely describe his model of fluvially eroded landscapes. And Stewart and Warntz (293) have discovered a dimensionless number, q:

$$q = \frac{P_T^2}{\int V^2\,dA} = 0.11$$

where P_T is total population, V is population potential, and A is area. The value of q has remained constant at about 0.11 for the 48 contiguous states in the USA since 1900. It has been determined that q is a general index of the distribution of a population; a small value of q indicates a central concentration of population within any areal unit; a medium value represents a roughly uniform distribution; and a large value, a boundary concentration (293, pp. 114–15).

The discovery of dimensionless numbers and the manipulation of dimensional formulae in the manner of the Reynolds number example above are more properly considered in the context of dimensional analysis, along with the question of how many dimensionless numbers may be expected in any system. Dimensional analysis will be examined in some detail, but more needs to be said about dimension first.

2. Dimensions of Length

In geographic inquiry, the major concern with dimensions has involved length. The familiar organization of the substance of geography in terms of points, lines, and areas (see chapter 4), reflects this emphasis. As we have seen in the previous chapter, a dimension of length is normally defined as '. . . n if an arbitrarily small piece of the space surrounding each point may be delimited by subsets of dimension $< n-1$' (159, pp. 10-24).[7] This means, simply, that the dimension of some phenomenon is one more than the dimension of the phenomena required to bound it. A line is bounded by points; areas are bounded by lines; and volumes are bounded by areas. Points, of zero dimension, are bounded by the empty set $\{-1\}$.

The dimension of length assigned to any object is in large measure a function of scale. Mandelbrot (202, p. 19) uses a ball of thread to illustrate this notion. At a distance of ten meters, the ball of thread may appear as a point, that is, as a zero-dimensional figure; at ten centimeters it is a ball, three dimensional; at ten millimeters it is composed of threads, one dimensional; at one-tenth of a millimeter each thread becomes a sort of column and the whole becomes a three dimensional figure again; at one-hundredth of a millimeter each column is resolved into fibers and the ball becomes one dimensional, and so on, with the dimension jumping repeatedly from one dimension to another. And below a certain level the ball of thread is represented by a finite number of atomic points, and it becomes zero dimensional again. Mandelbrot notes that the notion of a numerical result depending on the relation of object to observer is in the very spirit of physics of this century. The inevitable result is that depending on the criteria used, different observers may disagree as to the number of dimensions possessed by an object.

Note that the dimensions of length and the other primary physical and social quantities, although not necessarily the dimensions of derived quantities, are integers. Stewart (294, p. 31) discusses the occurrence of fractional exponents in dimensional formulae involving derivations of mass, length, and time, and cites 'mystical beliefs' about dimensions that hold that such exponents are 'unnatural'. He writes that fractional exponents are unsatisfactory not because of any mystical beliefs, but simply because they do not make sense physically. For example, just what would be meant by the square root (the one-half power) or the 1.4 power of a length? Stewart notes that if area were used as a primitive instead of length, fractional exponents would occur, but we know that area is not really a primitive.

Mandelbrot (202) expresses a contrasting view of fractional exponents of length. Whereas the topological dimensions of Euclidean space are represented by integer values (e.g. $[L^1]$ = length, $[L^2]$ = area, $[L^3]$ = volume), the Hausdorff-Besicovitch or fractal dimensions introduced by Mandelbrot need not be so. Coastlines, rivers, and the boundaries of snowflakes, which, although linear features, approximate to some degree plane-filling curves, may have fractal dimensions on the order of 1.2–1.3. These curves are in a twilight zone between one and two dimensions. The fractal dimension is demonstrated to be a useful measure of the complexity of linear features; the higher the dimension of a curve, the more complex it is.

Guyot (135) introduces into the geographic literature a more detailed consideration of the implications of dimension in length. The Meeting Theorem, which states the upper limit to the number of dimensions that figures might meet in, is related to the definition of dimension given above, where dimension is one more than that which bounds it. Meeting is defined as the contact in common in the form of a mutually inclusive figure of dimension n_c (called a meeting figure) between figures all of the same dimension n. That is, a figure of dimension n_c exists such that every part of it will be simultaneously common to m figures, all of the same n. Where:

n_c is the dimension of any meeting figure;

m is the number of figures all of the same n, that are meeting;

n_{c_m} is the maximum dimension of a meeting figure for m n-dimensional figures. This maximum dimension is the highest absolute value of n_c (135, p. 68).

Everywhere is the meeting figure common to all m of the n-dimensional figures. For example: two lines ($n = 1$) meet in a point ($n_{c_2} = 0$), two surfaces ($n = 2$) meet in a line ($n_{c_2} = 1$), and two volumes ($n = 3$) meet in a surface ($n_{c_2} = 2$). Therefore, when $m = 2$ the maximum dimensioned meeting figure between two n-dimensional figures is $n - 1$. This can be assumed since the highest dimension that is common to both n-figures is their $n - 1$ boundary. (This returns to the definition of dimension by the dimension of that which bounds; the meeting figure is the bounding object.) Further, one figure of dimension n meets itself in itself ($n_{c_1} = n$), and three n-dimensional figures meet in $n - 2$ dimensions. To summarize:

$$n_{c_1} = n$$
$$n_{c_2} = n - 1$$
$$n_{c_3} = n - 2 .$$

This may be generalized in the following way: $n_{c_m} = n - m + 1$ (135, p. 69).

Finally, Whitrow (354) takes a philosophical viewpoint in considering the important and general question of what is meant by the statement that the universe appears to have a certain number, three – no more and no less – of spatial dimensions. Since the mathematical discovery of a higher space, a clear-cut problem has emerged concerning the origin of the three-dimensional character of physical space. In spite of a variety of recent attempts to show that three-dimensionality is either a necessary attribute of our conception of physical space or is partly conventional and partly contingent, the problem cannot be considered as finally solved. Whitrow suggests that the dimensionality of the universe, particularly of the world, is partly contingent and partly necessary, since it could be inferred as the unique natural concomitant of certain other contingent characteristics associated with the evolution of the higher forms of terrestrial life, in particular of Man, the formulator of the problem (354, p. 129).

3. The Identification of Dimension in Geography

As suggested previously, dimension is recognized implicitly by most geographers when they organize their inquiry at the punctual, linear, and areal levels. In his work on experimental and theoretical central place, Bunge (55, chapter 6) became one of the first geographic researchers to make the role of dimension explicit through his representation of consumers, markets, and places as phenomena of varying dimensional characteristics. He illustrates these relationships with the aid of a 'centrality cube' (Figure 5.18A). Figures B, C, and D represent cross sectional slices of the cube from front to rear. The vast majority of central place theory has dealt with zero dimensional places, or points (Figure 5.18B). Figure E is the same as B, but with the possible cases keyed to letters to facilitate discussion. Here, Bunge identifies case C with the classic central place theory of Christaller as expanded by Losch; case A applies to the work of Berry and Garrison (26), in which market shape is not considered; Hotelling (155) deals with case E; and the spaceless case F is the realm of the economist. Cases B and D have not received attention.

Woldenberg (363) deals with dimension on a more abstract level, noting that systems of flow, whether organic, social-economic, or physical, must maintain certain relationships between their various components in order to function sufficiently well to survive. These relations are reflected in the dimensional and dimensionless quantities which characterize the system. In the context of a discussion of allometric and isometric growth, he clearly illustrates the role of dimension in determining the morphology of a system. Imagine a system in which one unit of surface area is required for each unit of volume. In an organic system, for example, this relationship might be necessitated by light, air, and food requirements. If the properties of the system remain constant, an increase in system size would cause the volume to exceed that which can be served by the increase in surface area (since volume increases as the third power of length, and area as the second power) and at that point the growth of the system must stop. In Table 5.3 an organism of this type is represented by a cube. Area exceeds volume until the side length of the cube is equal to six. In order for the

Spatial Structure (II) Morphology: Topology; Dimensionality

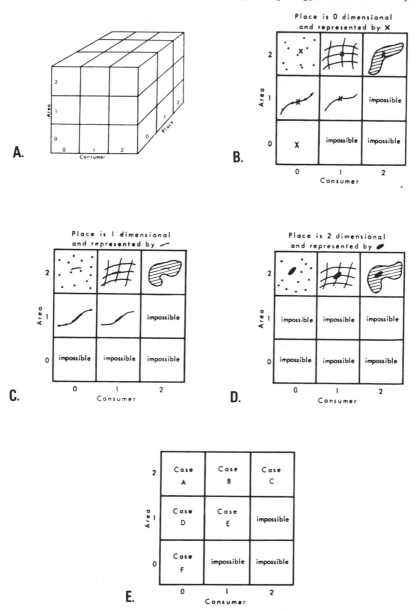

Figure 5.18 *Bunge's centrality cube*
    Source: Bunge (55, pp. 154–6)

organism to grow to a large size, area must be allowed to increase more
rapidly than volume (positive allometry). The basic dimensional relations
and, thus, the geometry of the organism would be altered; the organism
would no longer retain its cubic shape. Much of Woldenberg's work also
involves examining the dimensional correctness of allometric power
functions. This will be discussed further below.

Table 5.3 *Hypothetical organism in cubic form*

Side length	Area	Volume	Circumference
1	6	1	4
2	24	8	8
3	54	27	12
4	96	64	16
5	150	125	20
6	216	216	24
7	294	343	28

Coffey (79) and Dutton (95) extend the dimensional–allometric approach by examining the relationships between derived quantities (yielding dimensional rather than dimensionless numbers) and the manner in which these function as design criteria which influence the shape of various systems. The geometry of a system, such as the metropolitan and national systems considered by Coffey and Dutton, respectively, are found to remain relatively constant if the relations among the dimensions of the system, as expressed by the allometric power functions, are unchanged.

Other applications of dimensionality in geography have been in the field of dimensional analysis. We turn to this next, first reviewing the concept in general terms and then examining its utility in geographic inquiry.

B. DIMENSIONAL ANALYSIS

1. Definition

The second facet of dimensionality, dimensional analysis, is a tool to be utilized after the dimensions of the phenomena under investigation have been identified. Dimensional analysis, or the method of dimensions, has its origins in the 'principle of similitude' referred to by Newton in his *Principia*. Fourier's classic work, *La Théorie Analytique de la Chaleur*, published more than a hundred years after Newton's, expanded upon the method of dimensions and introduced two notions which are of primary importance in the modern content of the subject. The first of these, the dimensional formula, has been introduced above. As we have seen, corresponding to every physical quantity there may be ascribed a formula, known as its dimensional formula. This formula is based on the definition of the quantity in dimensional terms and shows the relation of any quantity to certain entities which are regarded as fundamental. Recall, for example, that the dimensional formula of population potential is $[NL^{-1}]$.

Fourier's second concept follows from the dimensional formula and is called the principle of dimensional homogeneity. Briefly, this principle states that the exponent of the dimension of any term in an equation must be the same as that in any other term. That is to say, the equation must be dimensionally homogeneous or dimensionally balanced. If it be not so, the inference is that an error has occurred in the analysis (158, p. 36). A word of caution must be added, however; an equation must be dimensionally balanced to be correct, but being balanced alone does not guarantee the validity of the equation. In general, then, dimensional analysis is a technique that has been used primarily in physics and engineering, but which has lately been fruitfully applied in geography. The technique may be utilized in deriving theoretical equations, checking empirical formulae, designing experiments, interpreting the results from scale models, and converting between different systems of units (151, p. 51).

2. Dimensional Analysis in Geography

Warntz (348, pp. 123–6) uses dimensional analysis to check the consistency of a land rent formula developed by Dunn (93):

$$R = E(p - a) - Efk$$

where R is rent per unit of land; k is distance; E is yield per unit of land; p is market price per unit of product; a is production cost per unit of product; and f is transportation cost per unit of distance per unit of product. The dimensional formula relating to this quantitative one is:

$$[DL^{-2}] = \{[ML^{-2}]([DM^{-1}] - [DM^{-1}])\} - [ML^{-2}][DL^{-1}][M^{-1}][L]$$

where D is value, L is length, and M is mass, as employed above.

In checking dimensional balance certain rules must be followed. Addition and subtraction are permitted only for phenomena of like dimensions, the result having the same dimensionality. Multiplication and division are permitted for phenomena of unlike dimensions, but the derived product or quotient assumes the resulting dimensional 'mix'. In multiplication, exponents are added for any similar dimensions; in division, exponents are subtracted. Raising to powers or extracting roots may be regarded as appropriately repeated multiplication or division. An operation that results in a dimension with a zero component cancels that dimension (348, p. 124). Following these rules, we see that the above equation reduces to

$$[DL^{-2}] = [DL^{-2}];$$

it is dimensionally consistent. Whether or not it is quantitatively correct depends upon the actual numbers involved.

Haynes (151) provides us with several examples of the use of dimensional analysis. In central place theory, the maximum distance anyone must travel to obtain a good is a function of its threshold in terms of sales per unit time, the average spending rate of the population on the good, and the density of the population. This function may be expressed as:

$$r = f(t, a, p)$$

where r is maximum travel distance, t is the threshold, a is the average spending on a good, and p is population density. Dimensionally, this function takes on the form:

$$[L] = [DT^{-1}]^x [DN^{-1}T^{-1}]^y [NL^{-2}]^z$$

where L, D, T, and N are the fundamental units employed above, and x, y, and z are exponents which need to be determined. Using dimensional analysis, Haynes demonstrates that

$$[L] = -2z$$
$$[T^0] = -x-y$$
$$[D^0] = x+y$$
$$[N^0] = -x+z.$$

Solving these simultaneous equations, he finds that the exponents x, y, and z have the values ½, $-$ ½, $-$ ½, respectively. Here, dimensional analysis has indicted the form of the relationship between the dependent and independent variables using no other information than the units of measurement of the variables.

In another illustration, Haynes examines Stewart's (292) original formulation of the gravity model as a force of demographic attraction, F, between two groups at points i and j:

$$F = G\frac{P_i P_j}{D_{ij}^2}$$

where G is a dimensional balancing constant analogous to the gravitational constant, P is population, and D is distance. Dimensionally, this equation is:

$$[MLT^{-2}] = \{[M^{-1} L^3 T^{-2}] [M] [ML^{-2}]\}$$

and is consistent or balanced:

$$[MLT^{-2}] = [MLT^{-2}].$$

A more recent version of the gravity model is expressed in terms of interaction, I, rather than the demographic force:

$$I_{ij} = k\frac{P_i P_j}{D_{ij}^b}$$

with the exponent of distance allowed to vary with each data set. Haynes notes that a problem with this formulation is that the dimensions of k depend entirely on the value of the distance exponent. Thus, changes in the exponent that occur with the calibration of the model in various situations enforce an inevitable change not only in the numerical values of k, which is not important, but also in the physical nature of k. Therefore, he suggests, serious doubt is cast upon the theoretical acceptability of the more recent formulation of the model. This is not strictly accurate, however, as the formula may be balanced by dividing by a constant:

$$I_{ij} = \frac{kP_i P_j}{(D/D_0)^b}.$$

Woldenberg (363) examined the dimensional consistency of the power functions which characterize various social and physical systems. He noted that if dimensional analysis establishes that a power function is not dimensionally balanced, it is safe to assume that allometric, rather than isometric, growth is occurring. During isometric growth, ratios between two parameters of unlike dimensionality would differ (e.g. length and area), but in contrast to allometric growth, ratios between two parameters of like dimensionality would be constant. During allometric growth, on the other hand, ratios between two parameters of like dimension would differ, while it is possible that the ratio between parameters of unlike dimensionality would be the same. At any rate, a power function relationship between two variables would not be dimensionally correct (363, p. 8). Further, Rosen (256) suggested that a dimensionally inconsistent power function is a reflection of system optimality. This notion of optimality will play a key role in the discussion of spatial process to be presented in later chapters.

3. The Buckingham Pi Theorem

It has been established that dimensionality includes the consideration of both substantial and natural variables or, in other words, dimensional and dimensionless quantities. The significance of dimensionless numbers has likewise been reviewed. Both the nature of dimensional relations and the existence of dimensionless numbers may be discovered through dimensional analysis. The manner in which dimensional analysis is employed in the derivation of the dimensionless Reynolds number has been illustrated.

In the dimensional analysis of a system, one of the most basic questions is that concerning the number of dimensionless quantities (conventionally referred to as pi's) that are needed to completely model the system. This question is answered by a theorem formulated by Edgar Buckingham of the US Bureau of Standards in 1914. This theorem gave the first clear statement and proof of the pi theorem that had been stated in a less general form by Riabovchinsky (294, p. 32). The Buckingham pi theorem provides a rule for deciding how many dimensionless numbers one may expect to find in any relationship; the number of independent dimensionless groups is equal to the difference between the number of variables that make them up and the number of independent dimensions involved. For example, if there are five variables involving three dimensions, two dimensionless groups would be expected. Or, in the fluid turbulence example, there are four variables and three independent dimensions and, thus, one dimensionless number (the Reynolds number). The weakness of the theorem is that it does not depend on the number of dimensions that have been actually used, but rather on the minimum number that might have been used (162, p. 171).

4. Dimensionality: Conclusion

An awareness of dimensionality, that is of the dimensions of physical and social variables, and of the dimensional consistency of formulae involving them, is requisite for a comprehensive understanding of the relationships between and within all terrestrial systems studied by geographers. Because of this, dimensionality is at once of great theoretical and practical utility to geographers and other scientists. In attempting to think about geographical problems from a dimensional point of view, we can learn something about the structure of spatial relationships and, more fundamentally, about the nature of the measurements used in geography (151, p. 64). As has been demonstrated, the legitimacy or illegitimacy of many of our conceptual structures and our quantitative models may be revealed in this way.

Finally, it is quite clear that dimensionality and morphology are not independent of one another; the dimensional relationships found to exist in a system directly and indirectly affect the form or structure of a system. Thus, dimensionality must be considered along with its coequal, morphology, as an integral component of spatial structure.

Notes

[1] See Adler (5), pp. 371–4, for a more detailed explanation of topological space.
[2] See ibid. pp. 99-100, for a discussion of the concept of orientability.
[3] A further example of this problem lies in attempting to develop a correspondence between thermodynamic phase space and the manifestation in physical space of those distributions of objects which appear to obey the laws of thermodynamics.
[4] Quoted in Warntz (339), p. 7.
[5] This demonstration was undertaken in an unpublished manuscript by William Warntz and William Coffey.
[6] See Ipsen (162), p. 42, for a more detailed explanation of the Reynolds number.
[7] The recursive definition of dimension may be traced originally to Poincaré (240).

6

Movement in Space

I. Introduction

The previous two chapters examined ways of viewing spatial structure. In the present one the movements which are both a cause and an effect of spatial structure are considered. It is in this sense of highly interrelated cause and effect that structure and movement may be regarded as dual constructs. Figure 6.1 illustrates this duality; it is the structure (6.1A) that determines the manner in which the movement (6.1B) takes place but, in a circular and cumulative fashion, the movement affects the structure.

Both Haggett (140, p. 31) and Crowe (81) have noted that the major emphasis on the part of geographers has traditionally been upon the static features on the earth's surface. In the preceding chapters, the discussion of these static features, structure, was relatively lengthy. This degree of detail was necessitated because the treatment of structure was at a more abstract and, thus, more generally applicable level than has been conventional in much of geographic inquiry. With a number of recent exceptions, many geographers have tended to examine specific configurations on the earth's surface at a very concrete level, ignoring the intrinsic structural properties underlying the distribution in favor of its unique substantive characteristics. For example, a given river system might be regarded in terms of its suitability for transportation and recreation rather than as one element in the general class of branching networks and topological trees. In contrast to the emphasis on the unique properties of structure, studies of movement over the earth's surface, although not widely undertaken until the contemporary period of geographic inquiry, have largely been of a more general nature, stressing models broadly applicable to many specific phenomena and situations; diffusion and migration studies are perhaps the prime examples of this approach.

It is for this reason that, unlike the discussion of structure, the treatment of movement will need to be relatively brief. Since many aspects of movement have been examined at an abstract level, and since they are quite well-known to a large number of geographers, there is no need to present a detailed explication of these topics. Rather, the purpose of this chapter is to attempt to organize these concepts into a useful and integrating framework. Movements occur in all facets of social and physical

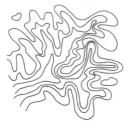

A. Contours of stream valley.   B. Movement of water down the valley sides.

Figure 6.1 *The duals of structure and movement*
   After Bunge (55, p. 260)

existence. Although there are certainly specific differences between various types of movement, particularly between movements involving physical and social phenomena, at an abstract level important homologies may be identified. It is the goal of this integrating framework to make such commonalities explicit.

All movement is a form of interaction. The term movement denotes the action of changing position. That is to say, it involves the progression of an object, idea, or energy from a point through other points, perhaps coming to rest at some destination point, either permanently or temporarily, or perhaps not. In this manner, two or more points will interact; some phenomenon either tangible or intangible is transported from at least one point to at least one other point. It is this interaction that makes social and physical reality as we know it possible; without it, 'existence' would be stagnation on a crystallized orb.

Before proceeding to a discussion of some of the concepts around which the examination of movement can be organized and thence to an appropriate framework in which to consider movement, it will be instructive for purposes of comparison to note the schema employed by Warntz and Bunge (345, chapter 12) in their treatment of movement. Quite briefly, their outline is this:

a.   The Earth's Motions. These include the earth's rotation, revolution, and other minor motions.
b.   Continental Drift.
c.   Erosion. This is used in a very broad sense. Diffusion, for example, is considered as a form of erosion. More will be said about this below.
d.   Circulations.
e.   Simulation of movement.

II. Movement: Organizing Concepts

The following concepts represent general considerations that need to be taken into account in the examination of any type of movement.

A. SCALE OF MOVEMENT

Whether or not motion is perceived to occur depends upon the spatial and temporal scales that are involved. Spatially, at the large scale (in the general sense of scale, as opposed to a map scale) distributions generally appear to be static. At a smaller scale, however, the movement of the elements of the distribution often may be perceived. For example, on a global scale (a small map scale), the population of New York City appears to be static. At the scale of the City itself, or of some section of the City (a large map scale), such movements as journeys-to-work, shopping trips, and recreational excursions may be discerned.

Similarly, temporal scale is an important consideration. Certain physical features such as mountains or continents, and certain social phenomena, a city for example, appear to be static when a small (short) time scale is employed. When a large scale is used, say 10,000 years for a continent or 30 years for a city, the movements inherent in these phenomena become apparent. The revolution of the earth around the sun is barely perceptible over a week but quite noticeable over six months.

The particular spatial and temporal scales employed must be specified, then, when discussing the movement or lack of movement of any terrestrial phenomenon.

B. MEDIUM, MODE, AND DIMENSION: CONSTRAINT

On the earth, movement may take place over land or water, through the land or water, or through the air. We may become even more specific by considering whether the land is a paved surface, gravel, sand or mud; whether the water is salt or fresh; whether the air is relatively near or far from the earth's surface. Each one of these media has a specific coefficient of friction associated with it that inhibits movement to some degree. In geography, the term 'friction of distance' is commonly employed. This usage is, of course, an incorrect one as friction is a property of the medium rather than of distance. Geographic space may be transformed, however, to represent the effect of the frictional properties of these media on time–distance or cost–distance (Figure 6.2).

The mode of movement is also of primary concern. The specific mode of movement affects the overall influence that the friction of a medium will have upon movement; a sandy beach presents more difficulties to a man on a bicycle than to one in a dune buggy; an upward sloping terrain presents more difficulties to the movement of a tree trunk by water than by glacier. (Note that water and glaciers may be simultaneously both modes and substances, that is they transport material and themselves as well; pp. 156–8.) In situations where the mode is inappropriate to the medium, the effect of friction may be so great as to prevent movement altogether; a conventional automobile is not able to move through deep water. Where movements in human societies are concerned, mode is clearly a function of available technology, as is the medium to a lesser extent. That is, man not only creates new methods of transporting himself, objects, and ideas but

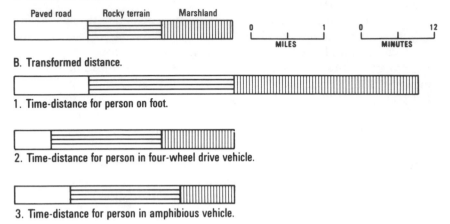

A. Untransformed distance.

Paved road   Rocky terrain   Marshland

MILES

MINUTES

B. Transformed distance.

1. Time-distance for person on foot.

2. Time-distance for person in four-wheel drive vehicle.

3. Time-distance for person in amphibious vehicle.

Figure 6.2 *Friction transformation of time–distance*

also modifies the media associated with these modes; he builds roads, bridges, tunnels and improved circuitry over which information may pass.

An additional and related concept is that of dimension. Depending upon the medium and the mode available, movement may be one dimensional, along a road or a canal, two dimensional, along land or water in any direction, or three dimensional, through land, air or water. Abler, Adams, and Gould (2, p. 239) use dimension to derive sixteen classes of movement (Figure 6.3).

Medium, mode, and dimension taken together yield the concept of relative constraint on movement. As demonstrated, medium and mode affect the time and cost expenditure of movement and, more fundamentally, whether or not movement is possible at all. This effect may be termed constraint upon movement, and one way in which it manifests itself is in the dimension of movement. For example, movement that is unconstrained over the earth's surface, say the movement of a man in an amphibious vehicle, is two dimensional; since the vehicle cannot fly, burrow (let us assume the absence of tunnels), or travel under the water, it is not able to move in three dimensions. Automobiles and trucks, in an urban environment, and trains are generally constrained to move one dimensionally, along lines, as are rivers in channels and telephone messages along wires (let us assume an intra-metropolitan telephone system where satellite relays are not necessary). Airplanes, submarines, and coal-mining machinery move in three dimensions. If the combination of the medium and mode is such that no movement is possible, the substance must stay at the point where it is located; movement is in zero dimensions.

C. SUBSTANCE OF MOVEMENT

The substance of movement refers, simply, to that which is moved or moving. Social movement generally involves organisms, information or

156

From (source or origin):	*To (destination):*			
	Point	Line	Area	Volume
Point	1	2	3	4
Line	5	6	7	8
Area	9	10	11	12
Volume	13	14	15	16

A. Classes of movement.

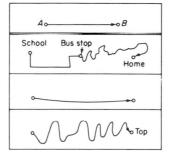

Ideal point to point movement.

Small boy returning home from school.

Path of a person attempting to walk directly across a football field.

Riding a horse to the top of a mountain.

B. Examples of class 1.

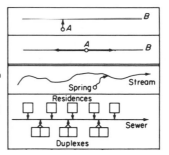

Ideal point to line movements:
From any point off the line to any point on the line.
From a point on the line to another point on the same line.

A spring (point) is joined to a stream (line) by a rivulet (another line).

Sewage lines collect from residential point sources.

C. Examples of class 2.

Figure 6.3 *Sixteen classes of movement defined by dimension*
    Source: Abler, Adams, and Gould (2, pp. 239–41)

ideas, goods, and money. In physical movement the substance may be air, water – in its liquid, gaseous, or solid states – soil, rock, pollen, continental plates, or even the earth itself. This is not, of course, an exhaustive catalog of substances. The particular substance that is moving will affect the ease with which the mode is able to transport it through the medium. For example, the wind is able to transport pollen with greater ease than it is able to transport pieces of granite. As we saw above, a substance may be self-propelling (that is, both mode and substance) as in the case of a man walking or a glacier flowing; a substance may also be medium, mode and substance as in the case of an ocean current. Substance together with medium and mode combine to yield the concept that Ullman (324) called transferability.[1]

D. VOLUME OF MOVEMENT

Closely related to the consideration of what is moving is the question of how much of the substance is moving. Clearly, the volume moved is related to the medium, the mode and the nature of the substance. A large swiftly flowing river is able to transport more sediment than a smaller, slower stream; the capacity to move bed load increases as about the third or fourth power of velocity.

More importantly, the volume may be employed as an indicator of the level of interaction or, more simply, the level of connectivity between two or more points. In the previous chapter we saw that a graph or its corresponding binary connectivity matrix describes spatial structure by specifying which nodes are connected directly or indirectly to which other nodes. As noted, the abstraction and generality of a topological representation is a trade-off against loss of information. One important piece of information that cannot be fully inferred from either a topological or a geometrical map is the degree to which two connected nodes interact. It is generally true, however, that connecting links arise because of the needs of the elements of a system to interact with one another; that is, structure is generally commensurate with the levels of interaction that occur. It is conceivable that two places may be linked by a road over which little or no traffic flows, however. The situation of East and West Berlin in the 1960s is an obvious example. In a case such as this, is it justifiable to consider these places as 'connected'? A knowledge of the volumes of movement is helpful when attempting to specify the degree of connectivity in a meaningful way. For this purpose, both directed and valued graphs are useful (Figure 6.4).

Unlike a conventional (unvalued, undirected) graph, directed and valued graphs provide a higher degree of information concerning the connectivity of nodes. The former indicates the direction of every edge (Figure 6.4B), while the latter displays the levels of interaction occurring (Figure 6.4C). A graph may be both directed and valued, providing more complete information. On the basis of this information we may distinguish between apparent and true connectivity. In Figure 6.4B node 4, although adjacent to node 1, is connected to the latter only through the intermediate steps of nodes 3 and 2. In Figure 6.4C there is no interaction between nodal pairs 1, 2 and 2, 3 even though links do exist. In these cases connectivity is apparent rather than real.

E. STRUCTURE OF MOVEMENT

The topics examined in the two preceding chapters are directly related to movement in that they define the structures over which movement occurs; as noted, these structures are the duals of movement. Thus, we may consider movement in terms of its geometry; the distance over which it

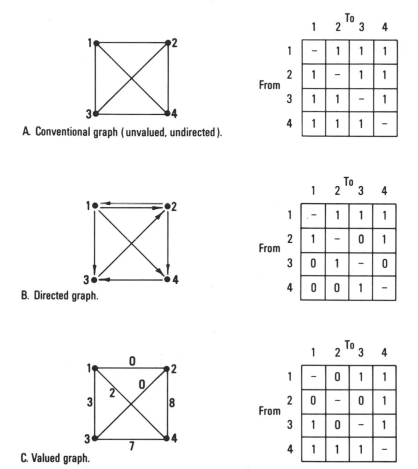

Figure 6.4 *Apparent and true connectivity*

takes place; the orientation of the movement; the shape defined by a movement or a series of movements; the pattern of nodes that are connected by movements; and the absolute and relative locations of the termini of movement and of the movement itself. In addition, the study of movement includes topological and dimensional consideration such as those touched upon in this chapter. Two related structural concepts involving movement may also be identified. These are the hierarchy of movement and the regionalization of movement.

Since structure and movement are duals, any structural hierarchy may also be considered in terms of a hierarchy of movement. Two of the most familiar examples of this are central place arrangements and diffusion. In central place arrangements, there corresponds to the structural hierarchy of places a hierarchy of movements that occur within the structure. Frequent shorter movements of people to lower order centres may be observed (Figure 6.5). In diffusion theory, some ideas and innovations

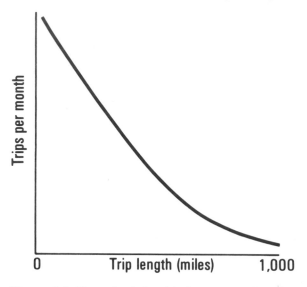

Figure 6.5 *General relationship between trip length and trip frequency in a central place system*

have been found to move hierarchically, leaping over intervening people and places as they are transmitted between larger places or important people first, and filtering down to lower elements of the hierarchy at a later date (Figure 6.6). In many instances, structural hierarchies are defined on the basis of the observed hierarchy of movement (Figure 6.7).

Two main approaches to the regionalization of movement may be identified in the geographic literature. The first of these involves the notion of a 'field'. A field may be defined as that area in which interaction (movement) with some node occurs. The sphere of influence of a city, the hinterland of a central place, or the migration field of a metropolis are all examples of fields. It is useful to distinguish between a field and a mean field. The former refers to the entire extent of movement, while the latter refers to that area that accounts for fifty percent of the interaction. For example, a small town newspaper may be sent to former residents living on several continents, an enormous field, but the majority of its readers may live within a radius of ten miles of the town. The size of a mean field has been demonstrated to be directly related to the transferability of the substance moved (140, p. 41). Thus, the mean field of a jewelry store may extend for hundreds of miles, while that of a brick factory may be no more than twenty-five miles.

The second approach to the regionalization of movement involves the delimitation of functional regions on the basis of similarities in the volumes of various types of movements. The primary tool employed in this type of study has been some sort of multivariate technique such as factor analysis or principal components analysis. The goal of these techniques is the 'identification of generic locational characteristics of groups of origins, or

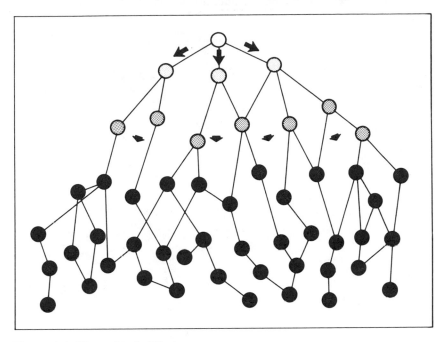

Figure 6.6 *Hierarchical diffusion*
Source: Abler, Adams, and Gould (2, p. 392)

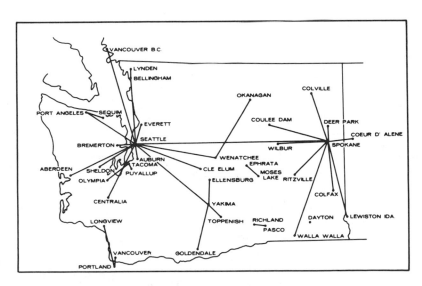

Figure 6.7 *Hierarchy of information flows: telephone calls*
After Nystuen and Dacey (224)

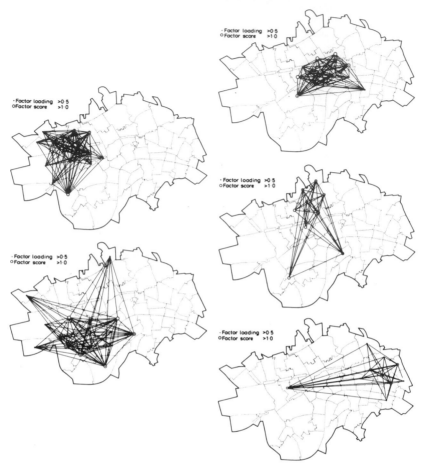

Figure 6.8 *Regionalization of taxi flows in London, UK*
  Source: Goddard (119, pp. 170–4)

groups of destinations, or of groups of origins and destinations (dyads)'
(279, p. 411). The pioneer study of this type was that performed by Berry
(23) on commodity flows in India. A recent contribution to this form of
analysis was Goddard's (119) examination of regions of taxi flows in
London, England. Using principal components analysis, he identified five
functional regions in the city (Figure 6.8).

For physical movement, hierarchies and regions may also be identified.
The hierarchy of streams is related both to the volume of stream discharge
(69, p. 232) and, as in the taxi example above, to the origin and destination
of individual flows. The various methods of assigning orders to streams
generally involve the consideration of how many lower order streams flow
into a particular higher order stream, and, in turn, how many streams of
that higher order flow into an even higher ordered stream (Figure 6.9). In
terms of regionalization of stream flow, the simplest method involves

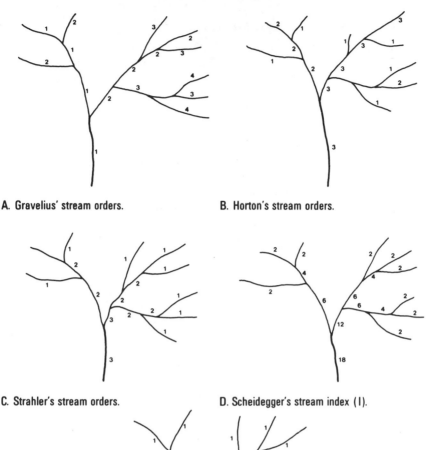

A. Gravelius' stream orders.

B. Horton's stream orders.

C. Strahler's stream orders.

D. Scheidegger's stream index (I).

E. Shreve's stream magnitude (M).

Figure 6.9 *Various systems of stream order hierarchies*

delimiting the drainage basins of streams. And as Warntz (340) has demonstrated, streams of various orders have drainage basins of particular orders in their own hierarchy associated with them. The existence of hierarchies and regions among atmospheric pressure flow systems has also been suggested (346; 299).

F. INTRINSIC CHARACTERISTICS OF MOVEMENT

The concepts examined thus far may be viewed as extrinsic ones in that they do not deal entirely with the inherent properties of movement but, rather, relate movement to the environment or frame of reference in which it takes place. Several intrinsic characteristics of movement may also be identified. These characteristics describe movement *per se* and not its external relations.

The first characteristic is that of directionality. Movement may be either unidirectional, one-way, or bidirectional, returning to some absolute or relative point of origin. The flow of a stream, for example, is unidirectional, while that of a tidal estuary, a journey-to-work, or the circulation of air or water is bidirectional. A second characteristic is continuity. In general, social movement is intermittent, involving frequent stops, while physical movement is often continuous. This distinction may likely have its basis in the distinction between the purposive behavior of human beings and animals, on the one hand, and the governance of physical phenomena by natural laws, on the other. Finally, there is the characteristic of permanence, which is directly related to the previous two concepts. Movement may be temporary in the sense of being either one component of a bidirectional flow or subject to pauses, such as a cross-country automobile tour. Movement may also be permanent; either a unidirectional flow of finite duration that never reverses itself, a migration, or a flow or circulation that is in motion permanently. Note that in ascribing any of these characteristics to movement, both spatial and temporal scale must be considered.

G. CAUSES OF MOVEMENT

Before moving on to the introduction of a general typology of movement, there is one additional concept of considerable importance that must be examined; this is the cause of movement.

1. Physical Movement

Physical movement, that movement in which man or animal plays no active role, depends upon some type of energy. The most important and most ubiquitous manifestation of this energy is the force of gravity. In the absence of strong opposing forces, all objects on or near the earth's surface are pulled toward the centre of the planet. An obvious example of this is the flow of water down slopes. Strictly speaking, of course, the flow of water down a slope, as opposed to its flow over the edge of a precipice, is not directly toward the earth's center but involves horizontal movement as well. As we have seen, the particular course that the water takes is determined by the contours of the landforms; the water moves in the manner of a freely rolling ball, orthogonal to the lines of equal height on the surface.

In atmospheric systems, too, movement is 'downhill'. In this case, the 'hill' is not a landform but a barometric pressure surface, and winds flow 'down' from the highs to the lows, moving orthogonal to the contours of the surface, the isobars. It is the non-uniform distribution of insolation and the differentials in the heating and cooling that are associated with it and with the relative locations of land and water that cause the non-uniform distributions of atmospheric pressure. If sunlight fell evenly everywhere upon the earth or, conversely, fell nowhere on earth, there would be no pressure differential and, therefore, no movement of air. Taking the movement of air an additional step, it is the drag of winds across the surface of water bodies that causes the horizontal movement of water.

Note that the orthogonal routes of water and air are nothing more than the geodesic paths that were introduced in chapter 4. Many other types of movement may also be ascribed to the influence of gravity; soil creep, earth flow, mudflow, landslides, glaciers, and lava are only a few that are readily apparent.

The force of gravity is counter-balanced by a number of other forces which tend to lift matter away from the center of the earth. One of these is convection, which involves some form of thermal energy for its impetus. Convection depends upon the fact that, in general, a rise in the temperature of a fluid results in a decrease in its density. Much of the upward movement of air is the result of the convection that occurs when insolation heats a parcel of air near the surface sufficiently to form a convection cell which rises much as hot air and smoke do in a chimney (299, p. 192). Similarly, continental drift is ascribed by some scientists to convectional currents in the earth's mantle (234, pp. 166–188; 327, pp. 233–6).

Another type of force that counteracts gravity is the capillary action that takes place within soil and vegetation; this results from the molecular attraction within the pore spaces in soil or plants (80, p. 550). Finally, there is hydrostatic pressure, which enables streams to flow uphill and artesian wells to pump water vertically. A stream course is not a monotonic decline; it contains short sections of upward gradients. This may be observed in the manner in which a stream dries up. It does not do so in a sheet-like fashion but first reduces to a series of disconnected pools (topological pits) formed by the downward and upward sloping of its bed (Figure 6.10). It is hydrostatic pressure which allows water to overcome these upward gradients.

Many physical movements represent an equilibrium between these counteracting tendencies; the downward pull of gravity is balanced by the thermal, molecular, or kinetic energy of the upward forces. The hydrological cycle is an obvious example of this equilibrium. Oversimplifying the cycle, vapor evaporated from bodies of water and due to the transpiration of plants undergoes convection as a direct result of the sun's energy. Eventually, the vapor condenses and falls back to earth, where the cycle begins again. And, as indicated, the flow of a stream is actually the result of both gravitational and hydrostatic influences.

In the above discussion of movement, one important consideration has

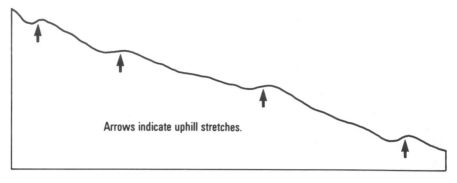

Arrows indicate uphill stretches.

Figure 6.10 *Profile of a hypothetical stream course*

been ignored. If the earth did not rotate, movements such as those of wind and water would occur in the manner described, orthogonal to the contours of their respective surfaces. The earth does rotate, however, and this produces another force, the Coriolis force, which tends to deflect moving objects. The effect of the Coriolis force is stated in Ferrel's law: any object or fluid moving horizontally in the northern hemisphere tends to be deflected to the right of its path of motion, regardless of the compass direction of the path; in the southern hemisphere, a similar deflection occurs toward the left of the path (299, p. 157). Objects at the equator that move parallel to the equator (parallel to the axis of rotation) are not deflected. The effect of the Coriolis force increases progressively toward the poles. Figure 6.11 illustrates the manner in which winds are deflected at various locations over the earth. The Coriolis force and the gradient force, which acts in the direction of the pressure gradient, generally reach a balance when the wind has been turned to the point at which it flows in a direction at right angles to the pressure gradient, parallel to the isobars. The wind in this state of balance with respect to the two forces is termed the geostrophic wind (Figure 6.12).

Ocean currents, too, are affected by the Coriolis force and circulate clockwise in the northern hemisphere and counter-clockwise in the southern hemisphere. Streams in the northern hemisphere occasionally show a tendency to undercut their right hand banks because of this force (299, p. 158). Bunge (345, chapter 12) has calculated that at the latitude of Detroit, Michigan, an automobile travelling at sixty miles per hour is deflected fifteen feet to the right for each mile that it travels.

2. Social Movement

Social movement, the active movement of human beings, animals, and insects, is a more complex phenomenon.[2] As in the case of physical movement some energy, whether provided by carbohydrates and glucose or by coal and oil, must ultimately underlie social movement. Neither energy nor the resultant force, however, are causal factors. In a majority of cases, the cause of social movement involves some purposive or teleologi-

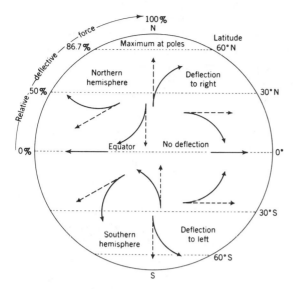

Figure 6.11 *Deflection of winds by Coriolis force*
Source: Strahler (229, p. 157)

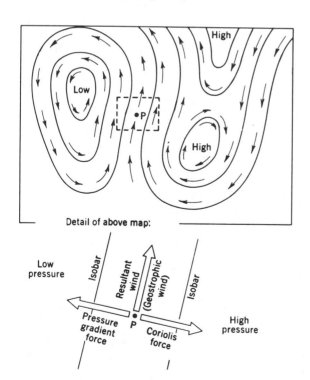

Figure 6.12 *Balance between the Coriolis force and the gradient force: the geostrophic wind*
Source: Strahler (229, p. 158)

cal behavior; in a smaller number of cases, primarily among animals and insects, some innate response to environmental and social conditions, the migration of lemmings to the sea, for example, is involved. Since innate response accounts for a relatively small proportion of social movement and since it is more in the realm of the ethologist, we can pass over it and concentrate upon purposiveness.

The causes of purposive social movement are a function of a complex set of individual and collective decisions. There are two viewpoints on this decision-making process. The first is that the decision procedure is strongly influenced by desires, attitudes, values, and perceptions concerning the environment and the relationship of the individual or the group to it. It follows that two persons may view the same environment in very different ways. This approach is the concern of behavioral and/or cultural researchers. Simmons (275) and Brown and Moore (49) have dealt with intra-urban migrations and have attributed the impetus to movement to a response to the stress that arises between the collective needs of a household and the perceived characteristics of its environment. Migration is an action that seeks to reduce this stress. Gould (126) formulates migration as an attempt to improve residential utility. The residential utility surfaces of individuals are elicited through mental maps.

The second model of decision-making involves the assumption of *homo oeconimus* and posits that on the basis of the objective characteristics of the environment an individual will make a choice that is optimal, that will maximize some utility or preference function. The implication of this model is that, having perfect knowledge of the true or objective characteristics of the environment, people and animals will move, insofar as they are able, to better satisfy their desires with respect to goods, information, services, food and so forth at some location other than their present one, to the extent that alternative locations are capable of satisfying such desires.

While still in the objective mode, we must note that there is a spatial separation, a quantitative and qualitative variation, of resources, in the broadest sense. That is to say, some places are more well endowed than others with regard to goods and services provided, food, fuel, raw materials, environmental quality and related amenities. Many of the disparities in the distributions of goods and services may be ascribed to the space economies of developed nations where activities are differentiated spatially because of the benefits of specialization, agglomeration and economies of scale (191, p. 2). The spatial variation in demand makes the problem even more complex; while production is often highly localized, consumers may be highly dispersed. Thus, it is the inequalities in the distribution of such things as goods, services, information, jobs, and people from place to place that necessitates interaction. In a social context, this interaction may be achieved through the movement of either the individual members of a society, of the commodities, or through a combination of the two.

Realistically, the causes of social movement involve both perceptual and objective factors. Spatial separation does exist and individuals often attempt to behave rationally, to minimize or maximize some function. The manner in which individuals ultimately act, however, is strongly influenced by perceptual and value-related factors. Note that the whole question of what happens after a decision to move is not considered here. For example, after an individual realizes that he needs food or needs to move to a region with more employment opportunities, how does he decide where to go? This matter is important but too complex to consider in the present overview of the general characteristics of movement. Similarly, the role of non-human social beings has been largely ignored here as it has been generally in social science. One may suggest, however, that animal and insect movements are based, to some extent, on simple perceptual and objective considerations.

In summary, social movement may be observed to be influenced by both positive and negative factors. The positive or pull factors include the attraction of an area where jobs or food are more abundant. In the same way, commodities and information flow to where there is a demand for them. The negative or push factors include things such as overcrowding, unemployment, predators, or a surplus supply of some commodity. Once again, in reality, movement results from both of these types of factors.

3. Social and Physical Movement: Commonalities

The primary distinction between physical and social movement is the presence or absence of purpose. Physical movement results from energy being imparted to some substance and, once the substance is in motion, it obeys the laws of physics. Social movement requires energy and obeys physical law but generally implies purposive behavior. The 'generally' is included in order to cover innate response of animals and insects. In spite of the many obvious dissimilarities between social and physical movement, there are certain important isomorphisms between them. These isomorphisms are based upon the lowest common denominator – physical law.

First, all movement is entirely dependent upon some form of energy and cannot occur without it. Once in motion, objects must obey physical law. An automobile or an ocean current deflects to the right in the northern hemisphere; a boulder on a mountain side or a man stepping out of an airplane is drawn toward the earth's center. Further, energy is manifested in some form of force. On the one hand, the force may be an attractive one such as gravity. The earth attracts all substances toward it, and a large city similarly attracts people, goods, and information (see the discussion of the gravity model below). On the other hand, the force may be a repulsive one. The heat of the earth's core repels the mantle from it causing convectional currents; an overly large human or animal population repels some of its members.

Second, both physical and social movement are generally characterized by minimum paths. That is to say, movements generally occur in such a

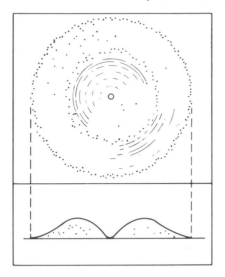

Figure 6.13 *Minimum effort in the construction of an ant hill*
  Source: Abler, Adams, and Gould (2, p. 237)

way as to minimize distance, energy, time, or cost. We have already noted the manner in which the movement of water, air, rocks and so forth down the gradients of their respective real or conceptual surfaces conforms to geodesic paths. Citing a further example that is more astrophysical than geographical, light rays follow minimum distance paths. Social movement is generally undertaken in such a way as to minimize the expenditure of time, cost, or energy. Even movement that is not ostensibly least effort, such as a cross-country sightseeing tour, involves minimization; it may attempt to maximize the sights seen while operating under time, money, and energy constraints. The route chosen will likely be an attempt at a geodesic specific to the situation and the constraints. Even the ant constructs its nest in such a way as to minimize the distance that soil must be transported (Figure 6.13).

One further illustration of the manner in which movements tend to follow the slope lines orthogonal to the isolines on surfaces involves maps of population and income potential. At the metropolitan level, it has been demonstrated (77; 78; 79) that the migration of people is generally down the gradients on the income and population potential surfaces (Figure 6.14). At the national level, studies by Stewart (291) and Warntz (336) indicate that the population is migrating up income and population potential surfaces (see Figure 3.1). The distinction between movements up and down the gradients on these conceptual surfaces seems somewhat paradoxical. One way in which this contradiction may perhaps be circumvented involves the sign of the potential quantities. In the examples above, potential is taken as a positive quantity. If potential is taken as a negative quantity, however, as it is conventionally treated in physics, the

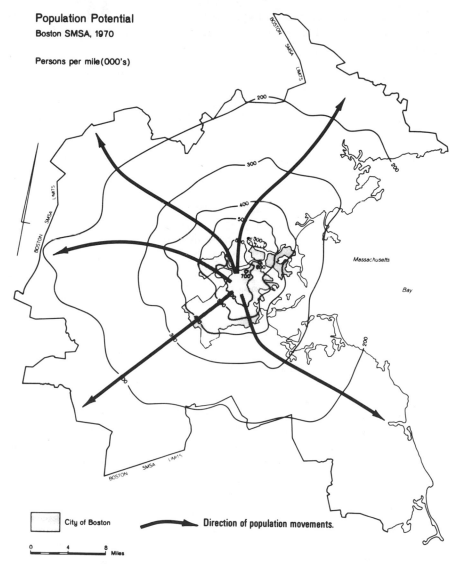

Figure 6.14 *Population movements down gradients on population potential surface*

peaks on these maps would become pits, and other complementary features would be interchanged (347, p. 181). If potential were taken as a negative quantity at the national level and as positive at the metropolitan level, all population movements would be down gradients. This may seem like a facile way in which to avoid the contradiction, but it may be that at certain scales or in certain circumstances a change in the sign of potential quantities is necessary.

Bunge (55, p. 28) utilizes the minimum effort principle, more specifically minimum cost, in his formulation of the 'shifting rule'. The shifting rule describes and predicts shifts (movements) in the location of such diverse phenomena as highways, rivers, shopping centers and central business districts. The rule is stated as follows:

> Where capacity increases require physical expansion, where this expansion cannot be in the vertical dimension, and where the near space is made more 'expensive' by the presence of the phenomena (*sic*) itself, a shift will probably occur to a new location as near to the old location as the area of induced 'expense' will allow.

Bunge illustrates the usefulness and generality of this rule, citing evidence of shifts in the Mississippi River, US Highway 99 in Washington, as well as the expansion of central business districts and volcanoes.

Finally, both physical and social movements may be observed to be entropic. As we saw in chapter 4, one of the usages of the entropy, that derived from statistical thermodynamics, may be employed as a measure of the level of organization of a distribution.[3] A system in which entropy is maximized is in its most disorganized state, at equilibrium. When entropy is minimized, that is when negentropy is maximized, a system is in its most highly organized state. The total information of a system, equated to its negentropy, may be calculated and the 'information statistic' utilized to describe the spatial distribution of the elements of the system (Figure 4.24). This usage of entropy is distinct from, yet related to, the manner in which it is employed in classical thermodynamics. Here, it is simply stated that in all systems there is a tendency toward maximum entropy; the amount of free energy in a system always dissipates into bound energy. High entropy is characteristic of a system in equilibrium, where bound energy is evenly distributed over the system. The entropic process is irrevocable (although in the statistical thermodynamics interpretation it is not, it being regarded as a definite, although small, possibility that a pile of ashes may heat a furnace); although the entropy of some component of a system may decrease, the entropy of the entire system must always increase. The law of entropy is a fundamental law of nature that is involved in every aspect of human behavior and physical existence (113, p. 3).

From both classical and statistical viewpoints, the movement of physical material, people, goods, and information can be viewed as part of the entropic process. Not only do these movements require free energy and result in bound energy, but the net result of them may be observed to be the disorganization (in the information theory sense) of the systems of which they are a part. The effect of the movements that we have considered thus far is to transport some substance from localized areas of surplus to areas of deficit; that is, to reduce the level of organization in the system. These movements are analogous to the heat transfer in a thermal system in which the tendency is toward equilibrium, a uniform temperature. This is not to say that there are not localized areas, as there may be in any system, where negentropy is increasing, or that there are not movements that maintain negentropy. Even though this may be occurring, the entropy of the entire system is increasing as the movements that feed the negentropy themselves convert free energy to bound energy.

The migration of the population in the USA was demonstrated by Warntz (336) to occur in a manner which will eventually (the year 2005 is forecast) bring about equilibrium of regional income distribution. While regional differences are disappearing (that is, while the entropy of the system is increasing) it is conceivable and even likely that within any one region entropy may be decreasing as people move from rural areas to the cities. Even this increase in statistical negentropy may be considered as entropic, however, if one considers that although equilibrium is not occurring spatially, it is occurring aspatially. That is to say, although people may be moving in a manner so as to cause the system to become more organized (the spatial view), the effect is probably to allocate more evenly the large numbers of opportunities and resources that exist in a metropolitan area to the corresponding numbers of people. Thus, opportunities and resources are distributed over the population, as opposed to over geographic space, in a manner approaching equilibrium.

To cite other examples, in any diffusion process information, people, or goods flow from localized sources to areas of deficit; these substances spread or diffuse over the surface until some sort of equilibrium condition is arrived at. The convection of air, water, and the earth's mantle is entropic in that heat is transferred from source regions to deficit regions in a manner that tends toward thermal equilibrium. The movement of air, water or soil down gradients occurs in a manner that levels or smooths the surfaces and their gradients.

4. Models of Movement

Models of movement may be broadly differentiated on the basis of whether they employ a deterministic or a probabilistic (stochastic) approach (191, p. 167). The distinction is, of course, that the deterministic approach is based upon mathematical certainty or known physical relationships, whereas the stochastic approach is used in cases where the nature of the relationships is not known and is expressed as probability statements. The deterministic or cause and effect models of movement include such relatively unsophisticated formulations as location quotients, shift techniques, and regression analysis.[4] Perhaps the most widely used deterministic model is the gravity model (see chapter 3). Included in the term gravity model is the potential model, which may be regarded as a multi-body gravity model. The gravity model will not be described in detail, as Olsson (227), Isard (163), and Carrothers (60) have presented comprehensive discussions of both the historical development of the model and some of the operational and theoretical problems associated with its use. The most common criticism of the model has been that it represents nothing more than a description of empirical regularities and has no real theoretical basis. Olsson (227, p. 49) notes, however, that many cases do exist for which it is possible to give the empirical regularities valid theoretical explanations. Any objections based on the lack of theoretical foundation notwithstanding, the fact remains that the gravity and potential models have shown a remarkable ability to describe and to predict, although not to explain, spatial interaction.

A recent utilization of the potential model was undertaken by Warntz in his examination of student flows in the United States. With each state taken, in turn, as the receiving state, the number of students attending school in that state from origins in each of the other states was correlated with the income potential contributed by the sending states. The results exhibit uniformly high correlation coefficients; student flows may be quite accurately predicted from a knowledge of the income potential contribution of each state to each other state (Table 6.1).

Physical movements may also be modelled by deterministic approaches. In problems dealing with slope development and river action (69, p. 289), certain known physical relationships can be used with empirical data to yield accurate descriptions and predictions. The work of Nye (222) on the movement of glaciers and ice sheets is a particularly good example of the deterministic approach.

Stochastic models of movement are characterized not by cause-and-effect-type statements, but by those in which the chance or probability of some occurrence is specified. A common stochastic approach to both social and physical movement is Markov chain analysis. Brown (48), Harvey (149), and Musham (215), for example, have used this approach to examine movements of human populations, and Chorley and Kennedy (69, pp. 185 ff.) and Chorafas (67) have used it in studies of river discharge. Simulation is perhaps the most familiar example of stochastic modelling. The so-called Monte Carlo method of simulation forms the basis of the widely employed diffusion model. The list of human and physical studies that utilize this approach is very long; a representative sample might include the work of Hagerstrand (138), Morrill (212), Brown (46; 47), Hudson (157), and Gould (128).

Finally, under the stochastic approach to movement models, there is the entropy maximizing model (see chapter 3). (Note that this usage of entropy is of the form derived from statistical thermodynamics or information theory. The statistical and classical interpretations of entropy confuse some authors, such as Lowe and Moryadas (191), who attempt to couch the entropy maximizing model in the classical form to some extent.) The entropy maximizing approach is primarily the product of Wilson (359) and involves the determination of the most likely state of an urban, regional, or national system. The most likely state of a system is that particular state for which the combinatorial probability is at a maximum; that is, the state that can be achieved in the highest number of ways. The distribution of journeys-to-work in a metropolitan area, for example, can be shown to exhibit a tendency toward maximum entropy, the most likely configuration of the system possible given certain constraints. It is of considerable interest that work by Wilson (359) has demonstrated the relationship

Table 6.1 *Regression equations relating student flows and income potential**

Alabama	$\log S =$	0.9070 +	1.2543 $\log U$	$R = 0.90700$
Arizona	$\log S =$	2.9544 +	0.9188 $\log U$	$R = 0.81983$
Arkansas	$\log S =$	1.3244 +	1.0252 $\log U$	$R = 0.86651$
California	$\log S =$	4.5491 +	0.9558 $\log U$	$R = 0.91567$
Colorado	$\log S =$	3.8667 +	0.8018 $\log U$	$R = 0.72425$
Connecticut	$\log S =$	2.0588 +	0.9809 $\log U$	$R = 0.94325$
Delaware	$\log S =$	−0.8109 +	1.0994 $\log U$	$R = 0.90243$
Florida	$\log S =$	2.6397 +	1.1508 $\log U$	$R = 0.90481$
Georgia	$\log S =$	1.6726 +	1.2030 $\log U$	$R = 0.91879$
Idaho	$\log S =$	1.6841 +	0.9158 $\log U$	$R = 0.70679$
Illinois	$\log S =$	3.9666 +	0.7964 $\log U$	$R = 0.85725$
Indiana	$\log S =$	3.4052 +	0.8847 $\log U$	$R = 0.92623$
Iowa	$\log S =$	2.6526 +	0.9579 $\log U$	$R = 0.83959$
Kansas	$\log S =$	2.6851 +	0.9056 $\log U$	$R = 0.80807$
Kentucky	$\log S =$	1.7367 +	1.1045 $\log U$	$R = 0.92780$
Louisiana	$\log S =$	2.2050 +	1.0172 $\log U$	$R = 0.91595$
Maine	$\log S =$	−0.4362 +	1.2846 $\log U$	$R = 0.84618$
D.C. and Md	$\log S =$	3.5014 +	0.8742 $\log U$	$R = 0.94331$
Massachusetts	$\log S =$	3.6742 +	0.9568 $\log U$	$R = 0.93588$
Michigan	$\log S =$	3.0706 +	0.9239 $\log U$	$R = 0.91414$
Minnesota	$\log S =$	2.6271 +	0.9169 $\log U$	$R = 0.72840$
Mississippi	$\log S =$	0.6946 +	1.1568 $\log U$	$R = 0.87823$
Missouri	$\log S =$	3.2361 +	0.9531 $\log U$	$R = 0.89664$
Montana	$\log S =$	1.6776 +	0.8696 $\log U$	$R = 0.66028$
Nebraska	$\log S =$	2.4151 +	0.9175 $\log U$	$R = 0.75015$
Nevada	$\log S =$	−0.2879 +	1.0348 $\log U$	$R = 0.82343$
New Hampshire	$\log S =$	1.3824 +	0.9662 $\log U$	$R = 0.86872$
New Jersey	$\log S =$	1.9014 +	0.9611 $\log U$	$R = 0.94080$
New Mexico	$\log S =$	2.1617 +	0.9008 $\log U$	$R = 0.84589$
New York	$\log S =$	3.5376 +	0.9545 $\log U$	$R = 0.90948$
North Carolina	$\log S =$	2.1356 +	1.1644 $\log U$	$R = 0.92702$
North Dakota	$\log S =$	0.8324 +	0.8453 $\log U$	$R = 0.63925$
Ohio	$\log S =$	3.0514 +	1.0033 $\log U$	$R = 0.90251$
Oklahoma	$\log S =$	2.6543 +	0.9000 $\log U$	$R = 0.87164$
Oregon	$\log S =$	2.2097 +	0.9784 $\log U$	$R = 0.73923$
Pennsylvania	$\log S =$	2.5485 +	1.0592 $\log U$	$R = 0.87645$
Rhode Island	$\log S =$	0.8256 +	1.1396 $\log U$	$R = 0.93112$
South Carolina	$\log S =$	0.5929 +	1.2593 $\log U$	$R = 0.90421$
South Dakota	$\log S =$	0.8041 +	0.9336 $\log U$	$R = 0.57013$
Tennessee	$\log S =$	2.3402 +	1.1424 $\log U$	$R = 0.90915$
Texas	$\log S =$	3.8758 +	0.8800 $\log U$	$R = 0.89955$
Utah	$\log S =$	3.0219 +	0.8828 $\log U$	$R = 0.69779$
Vermont	$\log S =$	0.2802 +	1.1686 $\log U$	$R = 0.87336$
Virginia	$\log S =$	2.1907 +	1.0677 $\log U$	$R = 0.94198$
Washington	$\log S =$	3.0054 +	0.9164 $\log U$	$R = 0.73860$
W. Virginia	$\log S =$	−0.1379 +	1.2584 $\log U$	$R = 0.86065$
Wisconsin	$\log S =$	2.6072 +	0.9816 $\log U$	$R = 0.87122$
Wyoming	$\log S =$	0.8538 +	0.9722 $\log U$	$R = 0.71928$

*For example, the number of students (S) from other states attending colleges and universities in Alabama may be estimated from the amount of income potential (U) contributed to Alabama by all other states.

between the entropy and gravity models. That is, the deterministic results of the gravity model appear to be isomorphic to those obtained within the entropy framework. Thus, the maximum likelihood principles which underlie the entropy model may also be viewed as a theoretical basis for the gravity model. In physical geography, Shreve (272) has taken a similar approach in the examination of the topological configurations of channel networks. His work involves the question of which configurations have the maximum likelihood of occurrence.

III. A Typology of Movement

Having considered some of the general characteristics of movement, it is now possible to construct a parsimonious yet useful framework with which to organize the diverse terrestrial movements that have been considered. Note that the framework that is proposed is but one of many that are possible. It is, however, the one that is considered to be of the highest utility, as it combines generality with the ability to discriminate between the major classes of movement.

A. THE MOTIONS OF THE EARTH

The revolution and rotation of the earth are perhaps the most fundamental of all terrestrial movements in that they are direct causes of such basic phenomena as time relations, including the day-night cycle and seasonality, and the distribution of climate types. Further, the rotation of the earth produces the Coriolis force and, therefore, plays a major role in determining the specific trajectory of any form of movement that occurs on the earth.

B. CIRCULATION

One general class of movement *on* the earth, as distinct from *of* the earth, is circulation. The revolution and rotation *of* the earth may be considered as circulations also, but from an astronomical rather than geographical perspective. As we have seen, circulation may be either vertical or horizontal, or both. In general, circulation may be regarded as indicative of the existence of a physical or social system. The substance of movement interacts with the various elements of the system *in an ordered and sequential way,* distributing or interchanging some quantity among them. Convectional circulations, for example, maintain their atmospheric, hydrological, or geological systems by dissipating thermal energy; the human journey-to-work maintains the social–economic system by enabling the exchange of factor services for income to take place; the seasonal migration of the arctic tern maintains a delicate ecological balance. Since movement and structure are duals, the patterns of circulation that are observed may be employed to identify the existence of distinct integrative systems and subsystems among a seemingly chaotic environment.

C. DIFFUSION

Circulations may be disaggregated into a series of component flows (101, p. 7). This implies that the distinction between circulation and flow is primarily a function of the scale employed and, therefore, somewhat spurious. Yet, one might suggest that the distinction is a useful one in that it differentiates between the recursive path of a substance through an ordered series of system elements and the locus-to-locus movements which may or may not be ordered and which may or may not delimit a system. That is to say, place-to-place movements sometimes indicate the existence of a system and the nature of its boundaries, but place-to-place movement may also be a random or *ad hoc* occurrence that does not reflect any real organization. The flows of income between areas of differential social well-being in a metropolis (51; 78; 79) may be regarded as part of the mechanism by which the metropolitan social–economic system maintains itself, but movements of people, air, water, information, and so forth may also occur in ways unconstrained by any systematic influence.

The term 'flow' is a common one in the geographic literature. It is argued, however, that its usage is restrictive in that a flow is simply the outward and visible manifestation of the more general phenomenon of diffusion. Diffusion, of course, involves the spread (place-to-place flow) of a substance, be it people, goods, disease, information, air, water, or whatever. The probabilistic nature of diffusion reinforces the above observation that a flow may or may not indicate the existence of a system. Experience has shown that diffusion generally occurs in an orderly or organized manner, but there exist definite probabilities for movements that depart from the overall organized structures that may be observed.

Warntz and Bunge (345, chapter 12) suggest that diffusion and most other forms of movement on the earth should be considered as special cases of the general phenomenon of erosion. They speak, for example, of the diffusion of information as the erosion of the ignorance surface, and of intra-regional migration as the erosion of rural population. It seems, however, that these authors have their general phenomena and specific cases reversed, and that erosion is better considered as a particular example of diffusion. That is, erosion in its physical sense, as distinct from Bunge's metaphorical usage, may be viewed as the diffusion of soil from one point to other points lying at lower altitudes. Similarly, migrations of people or animals, fluvial systems, and the movement of air down barometric pressure gradients, to cite just a few examples, are all instances of the diffusion of a substance. And, to repeat, this diffusion is generally manifest in flows which, in turn, may or may not delimit a system.

The notion that all movement, with the exception of the earth's motions and circulations (which might not be regarded as exceptions as they may be viewed as aggregations of flows), is a form of diffusion may seem a bit simplistic. The point is that diffusion is commonly acknowledged to involve the spread of some substance, and this is exactly what any form of movement accomplishes: the spread of people, water, air, information, money – the catalogue is virtually infinite – over the earth's surface. From an abstract viewpoint such as this, much of the exceptionalism in geography may be dismissed as artificial.

IV. Conclusion

Having examined spatial structure and movement in space independently, it is now possible to attempt the task of examining the synthesis of these duals: spatial process. Recall the discussion in chapter 1 of the question of whether the term 'spatial process' was a meaningful one, in view of the solely temporal implications that are generally attributed to 'process'. The conventional geographic approach which considers process as change over time alone was seen as an implicit acknowledgement of a Kantian view of time and space as independent absolutes. If the view that is now commonly accepted in science is adopted, that space and time are relative and inseparable, it follows that process must involve continuous change over both space and time. Further, just as researchers often focused on the temporal aspect of process, holding space out of the study, it is frequently convenient for the geographer to focus on the spatial aspect of process, holding the time element in abeyance. The examination of spatial process begins with a phenomenon that was briefly touched on as an example in chapter 1 – growth.

Notes

[1] Ullman's concept also included a consideration of cost.
[2] An interesting account of the dynamics of animal and insect societies is presented in Wilson's *Sociobiology* (361).
[3] For a good discussion of the various meanings of the term entropy, and of the manner in which these meanings are generally confused, see Marchand (204) and Georgescu-Roegen (113).
[4] See Lowe and Moryadas (191), pp. 167–8, for a description of these approaches.

PART THREE

SPATIAL PROCESS

7

Spatial Process (I) Growth

The concept of spatial process was introduced in chapter 1, along with an argument concerning the validity of the usage *spatial* process. Chapters 4 through 6 resolved spatial process into its two components, spatial structure and movement in space. The task before us now is a closer examination of the synthesis of structure and movement, spatial process.

In the discussion of the general concept of 'process' in chapter 1, growth was cited as one identifiable process. It is now suggested that while any number of processes may be distinguished on the basis of the non-spatial properties involved, it is not possible to enumerate a series of discrete spatial processes. Rather, it is more correct to speak of spatial process, a highly complex and integrated phenomenon, individual but related elements of which may be identified. In the remaining portion of this study, two such highly interrelated elements will be examined: growth and organization. This is not to suggest that these two concepts completely exhaust the subject of spatial process, but rather, that growth and organization are the major aspects of change in the spatial characteristics of any system. Further, the role of optimality and equilibrium in growth and organization will be examined.

I. Growth

As it is used here, growth may be viewed as a generic term that includes both positive and negative growth, increase and decline. The decrease in the size of a population or the contraction of the boundaries of a political unit are correctly considered as a negative growth. Virtually every object that has come under human scrutiny grows in some way and this growth has either direct or indirect spatial consequences. Growth is by no means a simple phenomenon, for the multitude of objects that grow do so in a variety of specific ways. Nevertheless, all forms of growth do have certain common characteristics. Boulding (41) has attempted to make these commonalities explicit in his threefold classification of growth phenomena: (1) simple growth (2) population growth, and (3) structural growth. These types of growth are not, of course, mutually exclusive, and the growth of any real phenomenon may involve all three. The classification proposed by Boulding is a useful point of departure but may be generalized further by reorganizing it in a manner that makes a distinction between absolute growth and relative growth.

A. ABSOLUTE GROWTH

Boulding's first two growth types may be included under the broader concept of absolute growth.

1. Simple Growth

Simple growth involves the growth or decline of a single variable or quantity by accretion or depletion. A road may be extended in length by ten miles, or the areal extent of Canada may decline by about 600,000 square miles if Quebec separates. The main problem in the analysis of simple growth is that of finding a 'law' of growth that will express the size of the growing variable as a function of *time* (41, p. 66; emphasis added). The most elementary case of simple growth is that of growth at a constant rate, the growth of a capital sum at a constant rate of interest, for example. Here, the growth function is the simple exponential form. Further familiar examples of the exponential growth function are, in its positive form, the law of Malthus, signifying the unlimited growth of a population whose birth rate is higher than its death rate and, in its negative form, the decay of radioactivity.

Continuous growth, especially at a constant rate, is relatively rare; all growth generally must run into declining rates. This is so simply because as growth proceeds conditions less and less favorable to growth are likely to be encountered. Virtually all empirical growth curves exhibit in the long run the logistic or s shape. Figure 1.1 depicts both positive and negative exponential growth and logistic growth.[1] While many 'laws' may be formulated to describe absolute growth, none make a pretense of attempting to explain it.

2. Population Growth

Boulding (41, p. 67) defines a population as an aggregation of disparate items or individuals, each one of which conforms to a given definition, retains its identity with the passage of time, and exists only during the finite interval. 'Birth' occurs when an item begins to conform to the definition which encloses the aggregation, and 'death' when the item ceases to conform to this definition. This population concept is a very general one and applies not only to human and animal populations, but to things such as automobiles, motion pictures, countries, ideas, and anything capable of being defined. The phenomenon considered, then, is not a homogeneous entity, but a population composed of discrete parts whose aggregate changes result in growth. Positive growth occurs when 'births' exceed 'deaths'; negative growth occurs when 'deaths' exceed 'births'; when 'births' equal 'deaths', there is no growth. Further, birth and death rates are regarded as a function of the age composition of the population. And age composition, in turn, involves the *time* elapsed since birth.

Both simple growth and population growth are forms of absolute growth as they deal with the accretion or depletion of some quantity over time. Absolute growth may, thus, also be referred to as growth-in-time. As in the examples of the positive growth of a road and the negative growth of Canada, absolute growth may have direct spatial consequences. The point is that it may, rather than must, have them. The growth of a human population, for example, may be viewed as having no spatial consequences other than the indirect one that, *ceteris paribus,* a large population will occupy more space than a smaller one.

Expansion diffusion may be considered as a form of absolute growth. As every geographer probably knows, expansion diffusion involves the spread of an idea, a disease, or some other communicable entity from individual to individual with the total number of 'knowers' or 'adopters' showing an increase over time. Implicit in this formulation is that the rate at which 'knowers' are 'born' exceeds the rate at which they reject the idea. A similar but more sophisticated approach is taken by Smith (280) in his discussion of the commonalities between plant dispersal, message diffusion, herding in animals, and the growth of cities. Growth in these types of systems involves the absolute growth of the information among the individuals comprising the population. In both expansion diffusion and Smith's formulation absolute growth does entail spatial consequences.

B. RELATIVE GROWTH

In contrast to absolute growth, which is characterized by accretion or depletion over time, relative growth involves time-independent changes in the spatial relationships of the elements of a system. Stated another way, relative growth involves differential morphological development. This is not to say that relative growth does not include accretion or depletion of some quantity but, rather, that the dominant characteristic of the growth of any complex structure is both internal and external morphological differentiation. Further, relative growth deals with long-run developmental changes rather than the more ephemeral compensatory changes in quantity associated with the minor perturbations of a system. That is, it is evolutionary in nature.

In order to initiate the discussion of relative growth and in order to illustrate in a very general way some of its features, a brief review of the five principles of structural growth formulated by Boulding (41) as a part of his attempt at a universally applicable 'general theory of systems' will be useful. These principles are:

a. The principle of nucleation. Any structure has a minimum size, its nucleus. Once a nucleus has been formed, it is not too difficult to determine the manner in which growth proceeds. The formation of the nucleus itself, however, presents many problems which are quite different from those concerning the growth of an already established structure. The nucleus does not have to be homogeneous with the structure that grows around it; a raindrop forms around a nucleus of dust. Consequently, heterogeneity or 'impurity' in an environment is a very important factor in explaining change. Further, because of the nucleation principle very small quantities often produce effects disproportionate to their size. In geography, the nucleation principle is implicit in much of locational theory as well as in such concepts as 'critical mass', growth poles, and diffusion. In the latter case, the nucleus may be regarded as isomorphic to an 'innovator'.

b. The principle of non-proportional change. As any structure grows, the proportions of its parts or its significant variables cannot remain constant. It is impossible to reproduce all of the characteristics of a structure in a scale model of a different size. To cite an obvious example, a metropolis is more than a large village; there is a basic change in the proportions of various functions with increasing size. This principle will be considered in more detail below under the theory of allometry.

c. The D'Arcy Thompson principle. At any moment the form of any object, organism, or organization is a result of its laws of growth up to that moment. Something which grows uniformly in all directions will be a sphere. Something which grows faster in one direction than in others will be long. Something which grows faster on one side than the other will spiral. Growth creates form but form limits growth. Boulding suggests that this mutual relationship between growth and form is perhaps the most essential key to the understanding of structural growth.

d. The carpenter principle. In building any large structure out of smaller parts, either the dimensions of the parts must be extremely accurate or there must be 'something like a carpenter or bricklayer' who can adjust the dimensions of the structure as it goes along. This principle which is more in evidence in the growth of social and economic, rather than biological phenomena, really involves entelechy or, in systems terms, homeostasis.

e. The principle of equal advantage. This principle governs the distribution of the substance of a structure among the various parts of the structure. The concept of advantage states that units will tend to flow toward locations of higher, and away from locations of lower, advantage. Advantage is an inverse function of the relative quantity of units. If there are differences in advantages to units in different locations, units will tend to move from the low-advantage locations where there are too many to the high-advantage locations where there are too few. In other words, a system will tend toward a maximum entropy. This is precisely what has been found in social–economic systems by Warntz (336) at the US national level and by Coffey (79) at the metropolitan level.

Many of these five principles will be encountered again, some of them under different names, in the detailed examination of relative growth that follows.

Although aware of the counter-arguments which were outlined in the first chapter of this book, it is strongly suggested that if geographers wish to learn more about any terrestrial phenomena, it is necessary to cross artificial disciplinary boundaries. In the case of growth, particularly relative or structural growth, there is much to be contributed to our understanding by biologists. As has been pointed out on more than one occasion, biology is an inherently spatial discipline that has developed a considerable number of spatial researchers. Further, as the work of such General Systems theorists as Bertalanffy, Boulding, and Rapoport has clearly demonstrated, many of the laws or relationships discovered in one discipline are fundamental and universal regularities that transcend fields of study. Certainly, living organisms do not grow in precisely the same manner as man-made or other inorganic phenomena. But although there are obvious differences there are also obvious and, one may argue, meaningful similarities which should not be regarded as simple and gratuitous analogies. The investigation of relative growth commences, then, with the theory of allometry.

1. Allometry

The discussion of allometry will begin with a very brief review of the major principles that underlie this theory. Anyone in need of more detailed explication may consult a number of more complete sources.[2] Broadly defined, allometry refers to the study of size and its consequences, and relates the differences in proportions of one component of a system to changes in either the absolute magnitude of the system or a second component of the system.

The principle of allometry dates at least as far back as Galileo (*c.* 1638) but received its first formal interpretation by biologists in the mid-nineteenth century (28, p. 164). Modern interest in allometry in biology and other fields is a direct consequence of Sir Julian Huxley's work in the 1930s. The literature of biology and of other fields (39; 229) contains numerous cases of documentation of the manner in which both organic and inorganic systems grow so as to yield a change in proportions.

The change in proportions or in the shape of a system is required by elementary geometry, specifically by the area-volume relationship. Note that this statement is to a large extent a 'shortcut' rationale and should not be taken to imply that size increase is the efficient cause of shape alteration (129, p. 588). If geometrical similarity is maintained with size increase, any series of objects will exhibit continually decreasing ratios of surface area to volume. Area varies as the second power of length, and volume varies as the third power. Constant area-to-volume ratios, an adaptive necessity for many organic and inorganic relationships, can only be maintained by altering shape.

At this point it will be useful to recall the model of the growth of a cube-shaped organism used to illustrate the concept of dimension in chapter 5 (Table 5.3). The area of the cube exceeds its volume until the length of the cube's side is equal to six. In order to obtain necessary surface area with size increase beyond that point the basic cubic shape must alter, becoming convoluted or 'bumpy'. In the simple model postulated, if the shape of the cube did not change in this manner growth would cease as the surface could not adequately act as an interface between the volume of the cube and the environment. Thus, differential increase is, in general, growth limiting at a certain point. We will return to this notion below.

The basic allometric relationship is described by a power function which has come to be known as the 'allometric equation':

$$y = bx^a$$

or its equivalent form,

$$\log y = \log b + a \log x$$

where x represents the size of the entire system, or a portion of that system that is being used as a frame of reference; y is the size of a particular element of that system; b is the y intercept when $\log x$ equals zero; and a is the exponent, the coefficient that relates the amount of change in $\log y$ per unit change in $\log x$.[3,4]

On the basis of the value assumed by the exponent, three categories of allometry can be distinguished. The first category is that of positive allometry. This means that y has a differentially large increase relative to x; the exponent a, which represents the ratio ($\log y/\log x$), is greater than one. The second is negative allometry. Here y decreases relative to the increase in the magnitude of x; the exponent is less than one. Finally, when x and y maintain a one-to-one correspondence throughout their increase, growth is said to be isometric; the exponent is equal to one.[5] One implication of the allometric equation is that the ratio of the increase in the y parameter relative to that of the x parameter is constant, the specific form of this constant relationship described by the exponent.

a. The Constant Ratio

There are several striking parallels between the concepts of allometry and entropy. First, each has wide applicability and together they form the cornerstone of general systems theory. Second, as is well-known in the case of entropy, and as will be discussed below with respect to allometry, each involves the notions of probability and maximum likelihood. Finally, there is considerable confusion and ambiguity concerning the usage of these terms. The preceding chapter touched upon some of the problems associated with interpreting the term 'entropy'; the sources of confusion regarding the principle of allometry, particularly its use in geography, will now be explored.

Three basic aspects of the principle of allometry may be identified. First, there is the existence of a constant ratio of the y parameter to the x parameter; this is indicated in the allometric equation by the exponent. For example, as the body weight of a fiddler crab increases, the weight of its large claw increases at a constant ratio of 8/5. Second, there is a change of shape with increasing magnitude, as exemplified by the simple model of the cube-shaped organism. Finally, in the allometric relationship, there is an implicit recognition of competition. The theory of allometry has enjoyed considerable popularity in geography in recent years. Geographers, however, with very few exceptions have utilized and, indeed, recognized only the first of these aspects. Each of the three will be reviewed in turn and its relationship to geographical inquiry made explicit.

The concern in geography with the existence of a constant relationship between two parameters of a system was preceded by the utilization of this principle in sociology. Pareto's law (229) of the distribution of income within a nation states that the number of individuals gaining a certain income is a constant proportion of the total amount of income. Similar empirical regularities concerning the population size of cities were observed by Auerbach (15) and Zipf (378) and articulated in the rank–size rule: the rank number of a city times its population is constant. Both Pareto's law and the rank–size rule may be stated as power functions having an exponent of negative one.

A comprehensive review of the applications of allometry in geography is not the purpose of this discussion but a few illustrations of the manner in which the constant ratio aspect has been emphasized will be instructive. One of the most common uses of allometry has involved relating the areal extent of a spatial unit to its population size. This was, incidentally, the first utilization of the principle of allometry in the geographic literature. Stewart and Warntz (293) found that for a sample of cities in the USA and Europe, the relationship

$$\text{area} = b \text{ population}^{3/4}$$

held between the years 1890 and 1951. Nordbeck (220; 221), postulating that all urban areas have the same shape and form, found one ratio between area and population in the built-up areas of Sweden ($a = 0.66$) and another in the Densely Inhabited Districts of Japan ($a = 0.91$). More generally, Tobler (317) has used satellite imagery to investigate the relationship between the same two parameters in ten countries. He found that the exponent values were specific to different cultures and suggested that the factor of proportionality, the b value in the $y = bx^a$ equation, may prove valuable as an index of settlement packing, reflecting strategies employed by various societies for the organization of spatial activity.

Woldenberg (363) has interpreted Warntz's (336) finding of the maintenance of an exponent of three in the power function relating income density and income potential as indicative of allometry in the US social–economic system. Similarly, Dutton (95) has found that the urban portion of the US population has increased as a constant ratio ($a = 1.73$) of the total population during the period 1790 to 1970. Other examples from human geography include Woldenberg's study (362) of the relationship between the total areal extent of an urbanized area and the size of areas taken up by specific land uses; Bon's (39) work on micro-environmental morphology which demonstrates, for example, that the ratio of the length of a communications network to the overall floor area inside a building is constant; and Bunge's (51; 56) interest in the relative proportions of the components of a city system. 'If you double the size of a city you double the size of the slums.'

Table 7.1 *Allometry in the income density (D) and income potential (U) relationship*

System		Equation		No. of cases	r
Canada	1971	$D = 7.48 \times 10^{-4}$	$U^{2.14}$	238	0.813
Toronto	1971	$D = 3.16 \times 10^{-14}$	$U^{2.26}$	447	0.757
Sydney	1971	$D =$ 0.03	$U^{2.89}$	40	0.972
Boston	1970	$D = 1.49 \times 10^{-17}$	$U^{2.65}$	536	0.826
USA	1970	$D = 2.90 \times 10^{-8}$	$U^{3.03}$	3068	0.735

In physical geography, allometry is manifest in the rank–size relationships of Horton's (154) law of stream numbers. The power function has also been used by a number of authors, including Morisawa (211), Strahler (297), Leopold (183), Leopold, Wolman, and Miller (186), and Woldenberg (367), to relate the measurable properties of a fluvial system such as length, basin area and perimeter, discharge, channel width, and meander length. Nordbeck (220) further demonstrates the relationship between the height and volume of a volcano.

Much of the investigation into the principle of allometry by Coffey (75; 77; 78; 79) has concentrated upon the constant ratio aspect in two types of systems, social–economic and fluvial. In the former, following Warntz, the nature of the relationship between income density and income potential in a number of countries at both the national and metropolitan scales has been examined. Table 7.1 summarizes the value of the exponent determined in several studies. The significance of particular values of the exponent is not yet fully determined, but the indication is that the exponent may represent a generalized measure of intensity of land use and of the homeogeneity of income distribution (77; 75).

Table 7.2 indicates some of the relationships found among various parameters in the Boston Standard Metropolitan Statistical Area (SMSA) in 1970. These power functions are posited to represent manifestations of certain necessary functional relationships within the system and a set of design criteria for growth in an urban system (79). Finally, Table 7.3 presents the results of investigations into the relationships among some of the parameters of 355 of the world's largest rivers.

Two observations follow from these examples. First, there are two distinct types of research that utilize allometry. One of these investigates the manner in which a single system develops; the other assumes that a series of individuals of the same class represent the various stages of growth of a single system. Dutton's study of the growth of the urban fraction of the US population and that of Warntz on the development of the US economic system are characteristic of the former, while much of the work on fluvial systems and intra-urban area/population and land use type relationships is characteristic of the latter. In the second type of study, the implicit assumption is that small rivers and small urbanized areas have the same form as, and represent the early stages of growth of, larger systems.

Table 7.2 Allometric power functions: Boston SMSA*

	Population P	Population density G	Income density D	Area A	Per capita income T	Total income I
P	—	-0.144 $G=12.2\times10^{4}P^{-0.343}$	-0.055	0.494 $A=8\times10^{-6}P^{1.34}$	0.328 $T=546.7P^{0.216}$	0.890 $I=546.7P^{1.216}$
G		—	0.961 $D=6874.7G^{0.920}$	-0.931 $A=7820.5G^{-1.06}$	-0.289 $T=6874.7G^{-0.08}$	-0.245 $I=5.38\times10^{7}G^{-0.141}$
D			—	-0.864 $A=2.5\times10^{7}D^{-1.03}$	-0.013	-0.047
A				—	0.375 $T=3511.5A^{0.091}$	0.543 $I=1.73\times10^{7}A^{0.273}$
T					—	0.723 $I=81.8T^{1.50}$
I						—

*Power function shown only where r is significant

Table 7.3 *Allometric relationships in the parameters of fluvial systems: the world's longest rivers*

Area	=	0.163	Length$^{1.92}$	$r = 0.898$
Discharge	=	3.79×10^{-3}	Area$^{1.01}$	$r = 0.707$
Discharge	=	1.09×10^{-3}	Length$^{1.85}$	$r = 0.606$

In their investigations into the growth of a particular species, biologists make the same assumption. The question is whether all cities or all rivers may be considered as members of the same 'species'. Does a small but mature fluvial system have the same form as one that is larger but less advanced? In summary, the second type of approach to allometry must be employed with caution by geographical researchers.

Second, many of the geographers that have utilized the allometric principle have applied the term 'allometry' to any relations that may be fitted by a power function. Nordbeck (220), for example, equates the term 'law of allometric growth' to 'power function' and then proceeds to use the power function only for relations exhibiting isometry, where there is no change of shape with size increase. Since allometry means size-correlated change of shape (this is implicit in its etymology), it seems that many geographers have been somewhat confused about its proper use. Many 'allometric' studies have focused on the aspatial statistical consequences of the exponent value but have ignored any direct spatial implications beyond oblique references to 'differing proportion'. Fluvial system researchers are to a large extent innocent of this aspatial bias, as the parameters with which they regularly deal, length and area for example, are basic geometric attributes.

b. Size-Correlated Shape Change

The term 'shape change' is a general one that includes alteration both of the external form of an object, the convolution of the surface of the cubic organism referred to above for example, and of the internal relations of the components of a system.

The most general statement that may be made concerning size correlated changes in shape involves the notion of dimensional equality. The power function

$$\text{volume} = b \text{ area}^{3/2}$$

is dimensionally balanced (the two side are dimensionally equivalent) since volume has the dimension $[L^3]$, area $[L^2]$, and $[L^3] = [L^2]^{3/2}$. Such a dimensional balance indicates isometry (see note 5). Therefore, when the two sides of an allometric equation do not reflect dimensional equality (that is, when there is either positive or negative allometry), the shape of an object cannot remain unchanged with increasing or decreasing size. For example, in a fluvial system one would expect to discover a set of dimensionally balanced power functions if an invariant geometric relationship were to be maintained with change in size. The relationships involving the world's largest rivers (Table 7.3) indicate that two of the power functions are clearly not dimensionally balanced, with the third (area and length) having an exponent close to the isometric exponent of two but possibly still significantly different.[6] The implication is that for the set of the world's largest rivers shape changes are related to variations in size.

In an urban example, if the areal extent of the central business district $[L^2]$ and the areal extent of the residential district $[L^2]$ are related by a power function whose exponent departs significantly from 1.0, the internal relations of the city must change. The exponent of 0.66 that Nordbeck (221) found relating the extent of built-up areas with their population sizes indicates no size-related shape change, as population may properly be regarded as a volume, since it is distributed in three dimensions. Thus,

$$\text{area} = b \text{ population}^{2/3}$$

is dimensionally balanced, but the relationship discovered by Stewart and Warntz (293) that

$$\text{area} = b \text{ population}^{3/4}$$

is not and implies that there is a change in shape associated with size variation in the sampled urban areas.

Woldenberg (366) utilized the findings of Leopold and Maddock (185) in his discussion of size-related shape changes in fluvial systems. The basic relationships between parameters discovered were:

$$W = aQ^b$$
$$D = cQ^f$$
$$V = kQ^m$$
$$(a)(c)(k) = 1$$
$$m + f + b = 1$$

where W is width, D is mean depth, V is velocity, and Q is discharge. Since m, f, and b are all less than one, width, depth, and velocity all grow more slowly than does discharge, Woldenberg observes (366, p. 2). Actually, this statement may not be accurate. The dimensions of the quantities involved are as follows:

$$W = [L]$$
$$D = [L]$$
$$V = [LT^{-1}]$$
$$Q = [L^3T^{-1}].$$

Ignoring the time dimension for simplicity's sake, we discover that the isometric forms of the power functions are:

$$W = aQ^{0.333}$$
$$D = cQ^{0.333}$$
$$V = kQ^{0.333}.$$

Since $b + f + m = 1$ is satisfied by these isometric values of the exponent, and since no one or two exponent values can be less than 0.333 without the remaining value(s) being greater than 0.333, it follows that width, depth, and velocity cannot *all* grow more slowly than discharge (that is, they cannot *all* have values lower than 0.333). Woldenberg continues that if, as is probably always the case, b does not equal f, then the shape of the channel changes with the increasing size of the system. This statement is accurate in that if at least one of the exponent values departs from 0.333, at least two of the relationships must be allometric and shape may be expected to change.

In similar work on fluvial systems, Hack (136) found that in a sample of fresh water streams

$$\text{length} = b \ \text{area}^{0.6}$$

indicating the progressive elongation of basin shape with increasing area. And Morisawa (211, p. 639) demonstrated that size-related change in shape occurs in the form of the more crenulated basin peripheries that are characteristic of large streams. In a further social illustration, Naroll and Bertalanffy (217) observe that the existence of allometry in the relation between the urban and rural proportions of the total population of a nation alters the form of the system in a way that indicates the force of attraction of cities toward the population as a whole.

A separate class of examples relating shape to size may be derived using the Pareto form of the allometric equation, the so-called rank–size rule. Recall that this 'rule' states that in a system there will be one element of the highest order, S of the next order, S2 of the next, and so on. In physical geography, this geometric progression is known as Horton's Law, and in human geography as the city rank–size rule. The specific form of the geometric progression is indicated by the bifurcation ratio. For example, a bifurcation ratio of 3 indicates that, for each order below the highest, the number of elements having that order number is a threefold increase over the preceeding number. Thus, the series would take this form:

Order (u)	Number of elements (N_u)
5 (highest)	1
4	3
3	9
2	21
1 (lowest)	81

The number of streams N_u having any order u is described by

$$N_u = R_b^{(k-u)}$$

where R_b is the bifurcation ratio, 3 here, and k is the highest order, 5 here. Both in river systems and in the branching systems associated with the human pulmonary artery, larger networks tend to have lower bifurcation ratios, probably because they have more difficult functions to perform (370, p. 25). This bifurcation ratio is related, in turn, to the shape of a drainage basin. A high bifurcation ratio is associated with an elongated

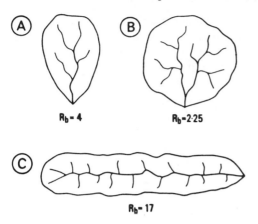

Figure 7.1 *Hypothetical basins of extreme and moderate bifurcation ratios After Strahler (299)*

basin, a lower ratio with a more compact basin (141, pp. 180–1). This is illustrated in Figure 7.1.

It is difficult to make a similar connection between the city rank-size rule and shape, as the rule is inherently aspatial. That is to say, unlike the fluvial systems in which order is determined on the basis of spatial information, which stream segments flow into which other stream segments, in the city case order is determined aspatially, solely on the basis of population size.

A final point concerning size-correlated shape change involves the notion of the limits to growth. Going back to the simple model of the cubic organism once again, recall that when the length of the side of the cube attained six units, the volume of the organism equalled its surface area. If the organism were to grow any larger, the volume would necessarily exceed surface area. Since this would surpass the ability of the organism to sustain life through interface with its environment, growth would have to cease when volume equalled area. As Boulding (41, p. 72) noted, growth creates form but form limits growth. The only manner in which the organism could continue to grow would be if the surface area increased disproportionately more than increase in volume; that is, if the exponent were lowered. Since

$$\text{volume} = b \text{ area}^{3/2}$$

indicates isometry, an exponent with a value lower than 3/2 would enable this disproportionate increase. And, as we have seen, this decrease in the exponent results in convolutions and bumps in the basic cubic shape.

c. Competition

The final aspect of allometry is one which, with the exception of Bertalanffy (28), has never been articulated. It has certainly not appeared explicitly in the geographic applications of allometry. Briefly, the allometric relationship may be viewed as an expression of competition within a given system, with each system component taking its share of the available resources of the total system according to its capacity, as expressed by the exponent. The exponent may thus be regarded as a 'growth partition coefficient' (160, p. 49) that expresses the capacity of a component to seize its share of the resources. An exponent indicating positive allometry, having a value greater than one when the equation is dimensionally balanced, signifies that the component in question captures a proportionately larger share of the resources than either the total system or a second component. Conversely, an exponent less than one indicates that a component captures a share proportionately less than the system or a second component.

The differential growth of various system components is, then, a consequence of the competition among these components for the resources available to the system from its environment. This is a direct and logical consequence of the theory of open systems, which concerns the continuous exchange of matter and energy between a system and its environment. In biology, the role of competition both among the components of an organism and among the species that comprise an ecological system has long been recognized. Bertalanffy (28, p. 66) writes that every whole is based upon the competition of its elements and presupposes the struggle between parts, and Thom (307, p. 323) suggests that all morphogenesis may be attributed to this type of conflict. One consequence of a sustained advantage in favor of one component is that in the long run the component or components with the smaller capacities will be exterminated. In the extreme case of growth, carcinogenesis, one component captures so much of the available resources that the total system not only grows proportionately less than the carcinoma but eventually declines in absolute size.

In biological instances of allometry, the resources of the environment that are the subject of competition generally include food, light, air, water and, not least, space. Indeed, Thom (307, p. 222) cites the competition for space as one of the most primitive (and therefore basic) forms of biological interaction, both among the internal components of an organism and among the components of an ecological system. In geography, the same basic form of competition may be identified. In fact, the arrangement of phenomena on the earth's surface may be generally conceptualized as the outcome of the competition for 'equipped' space, space having particular characteristics.

One geographic example that comes to mind involves Bunge's statements concerning the relationship of the size of the slum to the size of its 'host' city. By extension, this also implies a relationship between the slum and the other components of the city, the upper class residential district for example. Using several different criteria of size, we might expect to find, and to fit empirically the values of the two constants of, the following relationships between a city (C), its slum (S), and its upper class residential district (R):

$$\text{area S} = b \text{ area C}^p \qquad (1)$$
$$\text{population S} = d \text{ population C}^q \qquad (2)$$
$$\text{income S} = e \text{ income C}^r \qquad (3)$$
$$\text{area S} = f \text{ area R}^t \qquad (4)$$
$$\text{population S} = g \text{ population R}^u \qquad (5)$$
$$\text{income S} = h \text{ income R}^v. \qquad (6)$$

These equations not only describe the relationships between the slum, the upper class ghetto, and the whole city, but, if the values of the exponents are known, also describe the outcome of the competition for the resources of the urban environment – space, people, and income. Since all of the equations are dimensionally balanced, an exponent greater than one indicates that the slum is capturing a proportionately larger amount of the resources; an exponent less than one indicates that it is capturing a proportionately smaller amount; an exponent of one indicates that it is capturing its 'fair share'. On the basis of a stereotyped image of the urban system one might expect the exponent to exceed one in equations (1) and (2) and to be less than one in equations (3) and (6).

Other examples of competition that might be commonly found in a geographic context include the growth of a forest, with the conflict among different tree species for light, water, nutrients, and space, and the growth of a fluvial system. The latter case is less readily recognizable as competition as the equation

$$\text{area} = b \text{ length}^a$$

relates the geometric characteristics of a fluvial system rather than two discrete components. Nevertheless, it may be conceptualized as an expression of a conflict that partitions available energy into either a headward or a lateral effect. That is to say, it is a struggle for elongated versus compact space. In an aspatial example, Pareto's law of the distribution of income is clearly an instance of the competition for a scarce resource.

2. Allometry, Optimality, and Equilibrium

The allometric form of growth that has just been reviewed may be placed within a more general context of optimality and equilibrium considerations. The concept of optimality, that nature pursues economy in all her workings, is one of the oldest principles of theoretical science (256, p. 1). In order to consider optimality as anything but metaphysical, it must be recognized that the term optimal refers not to some ambiguous absolute but, rather, to an optimal solution among a set of very definite alternatives. Rosen (256, p. 2) suggests that in order to specify an optimal solution, one must: (1) determine the class of all possible solutions, (2) assign to each solution a number which represents a cost, and (3) search among the set of costs to find that which is the least. Thus, the task of finding an optimal solution may be properly regarded as one of selection; one solution is selected from the set of possible solutions to a well-defined problem.

In physics there are several well-known examples of the optimality principle. Fermat's Principle of Least Time states that a ray of light moving through an arbitrary medium whose refractive index may vary from point to point will follow, out of all possible paths, that path for which the transit time of the ray is a minimum. Maupertuis' Principle of Least Action asserts that a mechanical system, out of all possible paths consistent with the conservation of energy, will move along that path which minimizes the mechanical quantity known as action. Hamilton's Principle is similar to that of Maupertuis in that the possible paths of a mechanical system are consistent with the conservation of energy, but differs in that those paths that are considered are those which arise through virtual displacement (time is fixed and the space coordinates vary). Finally, there is the whole class of problems involving geodesics (see chapter 4); Plateau's problem of the form of soap films is one specific application in physics.

Although many physical systems may be observed to behave according to the principle of optimality, there is very little theoretical basis for this behavior. It is not known precisely how or why a ray of light comes to take the shortest path. Some of the geodesic problems in physics, Plateau's problem for example, may be explained through the tendency of a system to attain mechanical equilibrium. In biology, however, there is theoretical justification for optimality. This involves natural selection which, in turn, is a form of competition; those individuals or organs that are able to compete effectively will survive. Through selection biological systems tend to assume characteristics which are optimal with regard to certain circumstances (256, p. 6). On the basis of our knowledge of social and economic systems, there is no reason to assume that these systems lack a similar justification. A business that is not able to compete effectively in the market place will become 'extinct'.

Rosen (256, p. 7) puts forth the hypothesis that those biological structures which are optimal in the context of natural selection are also optimal in the sense of minimizing some cost function derived from the structural characteristics of the situation. More precisely, he suggests that the total cost of an organic system may be measured as the sum of metabolic energy (the intrinsic cost) and the pressure of selection (the extrinsic cost). This is known as the principle of optimal design and has a strong correspondence with Boulding's (41, p. 74) earlier and independent suggestion that biologists need to look for factors akin to cost and profit in order fully to understand growth.

Allometry may be conceptualized as a simple functional relationship between the sets of values assumed by the two systems components (form functionals, in Rosen's terminology) under consideration. This simple functional relationship which describes the nature of growth in a system represents the optimal alternative under a given set of conditions. This is demonstrated mathematically by Rosen (256, pp. 80–3). Further, Balkus (16, p. 6) notes that optimal forms are inextricably knit with conservation principles. Most systems that are in or are approaching an optimal state

196

behave in a manner that conserves time and energy; in a social–economic system income is likely conserved also. This is not to suggest that all systems which are characterized by allometric growth are in an optimal state, but rather that there is a tendency for them to approach that state.

The allometric relationship is also characteristic of open systems that are in a state of dynamic equilibrium or, put another way, that are in a steady state (69, p. 230). The steady state of an open system takes the place of, and is equivalent to, the equilibrium state of a closed system. Since this is so, there is a direct connection between allometry and probability. In a closed system, equilibrium, the state of maximum entropy, is the most probable state; recall that entropy is defined in terms of the logarithm of the probability of a state being attained (see chapters 4 and 6). Therefore, a system in a steady state, in dynamic equilibrium, may also be considered to be in its most probable state. In allometric growth a system conforms to a condition of maximum probability within the constraints of the given system (369, p. 125). Woldenberg (347, pp. 102–3) provides the mathematical derivation of the statement that allometry represents a maximally probable state subject to the constraints of flow within the system. Through the open system characteristic of equifinality, of course, the most probable state may be reached in a number of different ways, each beginning in any number of different improbable states.

3. Growth Gradients and Growth Centers

The allometric relationship which we have been considering, manifest in the power functions relating two components of a system, *describes* the manner in which growth occurs. It does not, however, *explain* the causes of relative growth. Recognizing that growth, whether of a biological, a physical, or a social system, has never fully been explained, the following exploration of some of the factors underlying differential growth is attempted. Although this attempt does not explain growth, it does isolate a number of more specific concepts. Growth will receive further consideration in the more general context of organization in the chapter that follows.

An understanding of the total pattern of growth in any kind of system may be approached by considering together the individual allometric relationships between the various components of the system, particularly the manner in which the growth partition coefficient varies spatially. Huxley (160) found that in biological systems the exponent values are generally distributed in an orderly fashion among the components (that is, the values exhibit spatial autocorrelation), with growth *gradients* radiating from *centers* of maximum differential growth. In biology little attention has been paid to the biochemical or physiological bases of these gradients, which Huxley (161, p. 468) suggests represent a graded distribution of some substance concerned with the regulation of growth.

In any growing system gradients representing the spatial variation in the intensity of growth or, alternatively, in growth potential, may be identified. These gradients normally have a single high point or growth center from which growth intensity grades downwards in both directions. In general, the less the difference between the growth partition coefficient of the component and that of the rest of the system, the flatter is the gradient with growth approaching an isometric mode. Any system appears to possess only a certain quantity of growth potential; if one part of a system is enlarged others must consequently be reduced (161, p. xi). This returns us to the notion of competition; the components of a system may be viewed as competing for a limited quantity of growth potential, those resources that collectively enable growth to occur. Finally, there is evidence that growth gradients change over time in response to shifting equilibrium conditions in a system during the course of its growth.

Gould (129, p. 626) suggests that it is (perhaps uniquely) in the realm of growth gradients and growth centers that the numerical values of the allometric parameters may mirror the actual efficient causes of growth.

Many examples of the existence of growth gradients in a geographic context come to mind. Newling's (219) rule of intra-urban allometric growth is one of these. Deduced from the proposition that population density declines exponentially with distance from the center of the city, the rule states that the rate of growth of population density is a positive exponential function of distance from the center of the city, expressed by the equation

$$(1 + r_d) = (1 + r_0) \, e^{gd}$$

where r_d is the percentage rate growth at distance d, r_0 is the percentage rate of growth at the center of the city, and g is the intra-urban growth gradient, measuring the rate of change of the rate of growth with distance from the center of the city. And since both density and the rate of growth are functions of distance from the center of the city, the rate of growth may be expressed as a function of density,

$$(1 + r_D) = AD^{-K}$$

where r_D is the percentage rate of growth during a given period when the density at the beginning of the period is D, A is a constant, and the exponent K is the ratio of the intra-urban growth gradient (g) to the population density gradient. Increasing density thus has a depressive effect upon the rate of growth. This relationship is made explicit in Newling's concept of critical density, a population density above which growth is negative and below which it is positive. Since density and distance are related, the critical density may be expressed spatially.

Ray (248; 249) provides several additional examples of the manifestations of growth gradients. The growth of railway systems in Europe may be seen to progress from west to east from the center of negative growth in Britain to the center of positive growth in the USSR. The growth of the French Canadian population in suburban Vancouver increases radially outward from a negative center in the heart of the French community. And the growth of urban centers in Canada in the twentieth century exhibits a gradient extending from east to west across the country, ranging from stagnation in the Atlantic Provinces to high positive increase in the Prairies and British Columbia. Ray suggests that the faster rates of growth on the periphery in all three systems is a compensation for the later start of the peripheral components. One might also suggest that it is related to the tendency toward equilibrium within the respective systems.

Growth gradients may be viewed as the manifestation of the spread effects of growth centers, both positive and negative. A center of positive growth induces positive growth in the surrounding area; one of negative growth induces relative decline. Note the obvious correspondence between the concept of the growth center and the growth pole concept of the French regional economist François Perroux. In most instances the growth pole is an artificially created growth center intended to stimulate its region. In the growth pole analogy, however, growth is often interpreted in an aspatial economic sense.

The general concepts of growth gradients and growth centers have implicit in them, and subsume, many of the more specific 'causes' of growth that are sometimes identified, including so-called (temporal) process laws and behavioral considerations. It seems that an analogy with biology is once again justified. The general concepts of growth outlined here are of a level of abstraction similar to that employed by theoretical biologists in their investigations into the qualitative and quantitative spatial aspects of morphogenesis. On the other hand, the concern in geography with individual preference functions and 'the process of adjustment in the housing market' is similar to the biochemical approach to examining morphogenesis, quite possibly of limited utility in understanding the larger phenomenon.

4. Conclusion

Fundamentally, all form is the result of differential growth (248, p. 3). This is not to deny the existence of kinds of relative growth beyond that described by the allometric power function, kinds in which the co-relation of two components is not described by a straight line but, rather, may be curvilinear or subject to periodic fluctuations or sudden jumps. These kinds of growth have not been dealt with here since the emphasis is upon the identification of general relationships. A new and promising method of treating some of these kinds of growth in a general manner, however, has recently appeared in Thom's *Structural Stability and Morphogenesis* (307) in the form of catastrophe theory.

Finally, a discussion of growth would not be complete without at least a brief word on values. In the society in which we live, the prevailing attitude is that growth is good, the bigger the better; our economy must grow, our cities must expand, urbanization must come to the remote developing regions. This set of values which equates the amount of good to society with the amount of growth is essentially a quantitative view that ignores the more important qualitative view. Simply stated, the qualitative view is concerned not with the quantity of growth but with the nature or kind of growth. The values of our society are such that growth is regarded as desirable, and the question of whether or not the growth may be malignant is rarely considered.

Notes

[1] The equations that describe these curves are:

$$\text{exponential growth: } Q = Q_0 e^{a_1 t}$$

where Q is the number of elements; Q_0 is the number of elements at time t_0; t is the elapsed time; and a_1 is a constant taken from a Taylor power series.

$$\text{logistic growth: } Q = \frac{a_1 C e^{a_1 t}}{1 - a_{11} C e^{a_1 t}}$$

where the variables are as before, with a_{11} being a second constant from a Taylor series, and C being the sum of the Taylor series terms $a_{11} + a_{22}$. See Bertalanffy (28, pp. 60–3) for a derivation of these equations.

[2] More detailed discussions of allometry are found in Gould (129), Woldenberg (363), and Bertalanffy (28).

[3] The power function equation may be derived in this manner:

$$F(dy, y, dt) = ay$$
$$F(dx, x, dt) = ax$$

and since the growth functions of y and x can be said to cancel algebraically,

$$\frac{dy}{y\,dt} \Big/ \frac{dx}{x\,dt} = a$$

$$\int \frac{dy}{y\,dt} = a \int \frac{dx}{x\,dt}$$

and eliminating dt from both sides,

$$\log_e y = a \log_e x + \log_e b$$
$$y = bx^a.$$

[4] Some authors, including D'Arcy Thompson (309), argue that the use of the logarithmic power function is inappropriate and that many of the trends described by it are equally well rendered by linear regressions. The power function, however, has been used almost exclusively as it combines an adequate statistical fit with simplicity and interpretability (129, p. 596). Also, since growth is multiplicative in the general sense that what is produced by growth is itself normally capable of growing, it is reasonable to compare growth on a logarithmic scale where addition of units represents a multiplicative effect (160, p. 281).

[5] This definition of positive and negative allometry and isometry is valid only when the x and y parameters have the same dimensionality. When the dimensions of x and y are not equivalent, a scaling adjustment must be made before the form of growth may be identified. For example, when y is an area $[L^2]$ and x is a volume $[L^3]$ an exponent of 2/3 is indicative of isometry.

[6] The null hypothesis that the exponents of two power functions are not significantly different may be tested using a t-test of the following form:

$$t_{n_1+n_2} = \frac{|b_1 - b_2|}{\sqrt{[n_1 S_{b_1}^2 + n_2 S_{b_2}^2/(n_1 + n_2)]}}$$

where b_1 and b_2 are the two exponents, n_1 and n_2 are the number of cases, and S_{b_1} and S_{b_2} are the standard errors of estimate of the regression coefficients.

8

Spatial Process (II) Organization

II. Organization

The second of the two components of spatial process is organization. Organization and growth are closely interrelated, and a cogent argument may be advanced for considering growth within the more general framework of organization. In the course of the examination of the organization of spatial systems that follows, some of the explanatory factors relating to the 'how and why' of growth that were given incomplete treatment in the preceding chapter will be made more explicit.

As has been the case with many of the concepts that have been dealt with in this discussion, organization is a complex construct. In order to reduce the ambiguity inherent in the usage of this term, several points need clarification. First, etymologically, organization refers to the qualities that characterize a living being, an *organism:* connection and coordination of parts, a systematic and orderly structure, and functional interdependence. This is not to imply that only living systems possess these characteristics, for General Systems Theory has demonstrated that a wide range of inorganic systems also manifest these attributes (see chapter 1). Rather, these qualities were first recognized in, and more universally apply to, organic systems.

Second, when employing the term organization, a distinction must be made between the condition of *being organized,* that is of possessing a structure or arrangement that is orderly or 'systematic' (this sense more correctly falls within the bounds of morphology; see chapters 4 and 5) and the process of *becoming organized.* It is the latter usage that is the concern of the present chapter.

Third, as with the concept of growth, organization is a generic construct. That is to say, organization may be both positive and negative; a spatial system may become more organized, gaining coordination of parts and interdependence, or it may become less organized.

Fourth, a useful distinction may be made between self-organizing and externally organized systems. This distinction represents a fundamental qualitative difference in the nature of spatial systems. Through the internal relations of systems components and the external influences of the environment, self-organizing systems undergo the process of *becoming* organized (either positively or negatively). On the other hand, externally organized systems *are,* rather than *become,* organized by an extrinsic control; an organized structure is imposed upon them. Phrased in another manner, self-organizing systems organize 'naturally', as opposed to the 'artificial' or 'supernatural' organization of externally organized systems (230, p. 162).

Finally, any consideration of organization, in the sense of both the existence of an organized structure and the process of organization that creates that structure, raises the epistemological and ontological question of whether order is an attribute of the real world or, rather, a function of human mental processes. One of the most fundamental human activities is the ordering of experience, the simplification of complexity. In fact, this is not an inappropriate definition of the human activity known as Science. If nature abhors the void, the mind abhors what is meaningless (173, p. 82). Show a person an ink-blot and he will attempt to organize it into discernible shapes and figures. When the Babylonians began to chart the stars, they first of all grouped them together into constellations (constellation literally means putting stars together) of lions, virgins, archers, and scorpions. Similarly, the Greek astronomers broke up homogeneous space into the hierarchy of eight heavenly spheres, each equipped with its clockwork of epicycles (173, p. 83).

Gerard (114, pp. 218–19) has coined a nice term for man's ordering of experience: entitation, the identification of entity. The degree of organization that is perceived in the real world is a function of the kind of entities that are employed in the attempt to structure reality. It is Gerard's feeling that entitation is more fundamental and vastly more important than 'quantitation'. The role of entitation in human intellectual activity is put in its proper (important) perspective by an ancedote that he tells concerning a discussion by three baseball umpires (114, p. 219):

> The first one said with some satisfaction, 'Balls and strikes, I call them as I see them.' The second, a little more arrogant, said, 'Balls and strikes, I call them as they *are.'* The third one, of greater experience and wisdom, said, 'Balls and strikes, they ain't nothing *until* I call them.'

To this notion of entitation may be appended Pattee's (230, p. 170) observation that if there are complex systems that are not organized (one might add 'in conventional ways'), they may to a considerable extent escape our observation and understanding.

A. ENTROPY, ORGANIZATION, AND ORDER

In chapter 4, and again in chapter 6, the concepts of entropy and information theory were introduced, and the distinction between them and the confusion concerning their proper usages pointed out. In order to develop clearly the major theme of the present chapter, organization as spatial process, it is necessary to utilize the related constructs of entropy and order. Before proceeding to the statement of this theme, let us examine the multi-faceted notion of entropy in a bit more detail.

Spatial Process (II) Organization

1. Entropy in Classical Thermodynamics

The concept of entropy originated in the branch of physics known as thermodynamics which, in turn, developed from the work of Carnot (*c.* 1820) on the efficiency of steam engines. One result of Carnot's investigations was the formal recognition by physics of an elementary fact known for ages: heat always moves by itself from hotter to colder bodies. Subsequent discoveries indicated that all known forms of energy, too, move in a unique direction from a higher to a lower level. By 1865 Clausius had formulated the first two principles of thermodynamics: (1) Conservation of Energy – the energy of the universe and of any closed system remains constant; and (2) Degradation of Energy – in the universe and in any closed system there is a tendency for the free (available) energy of the system to dissipate into bound (irretrievable) energy. Stated another way, the latter principle, also referred to as the Second Law, specifies that any system without inner constraint tends over time toward an irrevocable state of equilibrium or perfect homogeneity (disorder) in which it is unable to produce work. This is a state of maximum 'entropy', a term coined by Clausius from a Greek word equivalent in meaning to 'evolution' (113, p. 130).

The converse of maximum entropy is maximum negative entropy or 'negentropy' (minimum entropy, since negentropy = −entropy). Negentropy is maximized when a system is in a state of perfect heterogeneity (perfect order); in this state a system is able to produce the greatest quantity of work. Note that the ability of a system to produce work is not a function of its energy since, according to the First Principle, the energy of a system is constant. Rather, the ability to produce work is a function of the degree of negentropy of the system – whether it tends toward a maximum or a minimum, the way in which free energy is distributed throughout the system, in other words. A perfectly homogeneous system, for example one with a uniform temperature distribution, cannot produce work as its negentropy is at a minimum. On the other hand, a system that is perfectly heterogeneous (ordered) possesses a maximum of free energy and can produce the greatest quantity of work.

Reiterating the discussion thus far, the classical thermodynamic usage of entropy refers to the difference in the nature and distribution of two qualitative types of energy (free and bound) within a system. In the beginning, the chemical energy of a piece of coal in a furnace is entirely free; it is available for producing mechanical work. Eventually, the free energy dissipates completely throughout the system and becomes bound energy, energy which can no longer be used to produce work. The piece of coal becomes a pile of ashes. Thus, entropy is an index of the relative amount of bound energy in a closed system. More precisely, it is an index of how the energy of a system is distributed. *High* entropy means that most of the energy of a system is bound or unavailable (homogeneously distributed); *low* entropy means that the opposite is true; most of the energy is free (heterogeneously distributed).

Two related points need to be stressed. First, although the general tendency of a closed system is toward high entropy, individual components of a system may increase their negentropy or acquire a more heterogeneous distribution of energy (become more ordered). To use a simple analogy, a desk may become more ordered by creating further disorder in the rest of the office. Second, the irreversible or irrevocable nature of the degradation of energy led Eddington (99) to label entropy 'the arrow of time'. It is the irreversibility of physical events, expressed by the entropy function, which gives time its direction. In a universe of completely reversible processes, that is, in a universe without entropy, there would be no difference between past and future. Entropy functions do not, however, contain time explicitly (28, p. 151).

Before concluding this brief examination of the classical thermodynamic interpretation of entropy there is one additional point to consider which will be of relevance to the subsequent discussion of the organization process. The Second Law applies only to those systems that are closed, systems that do not exchange matter or energy with their environments. In other words, energy is neither added to nor substracted from these systems. Much of the real world, however, is composed of systems that are open. Here, energy crosses the system boundary in the form of a flow to or from the environment. Every living system is an open system, as are spatial systems such as rivers (184) and cities. Obviously, the conventional formulations of classical physics are, in principle, inapplicable to this class of systems. In fact, many characteristics of the nature of open systems appear to contradict the laws of physics.

This apparent contradiction is resolved in part through a relatively recent development in science, the theory of open systems. Under certain conditions, open systems approach a time-independent state, the so-called steady state. The steady state is maintained at a distance from true equilibrium and the system is, therefore, capable of producing work. In spite of continuous irreversible processes, the import and export of energy, and the building-up and breaking-down taking place, the system may remain constant in its composition (28, p. 142). While the change of entropy in closed systems is always positive (order is always destroyed), open systems are characterized not only by the production of entropy due to irreversible processes but also by the import of entropy, which may well be negative, from the environment. Thus, open systems maintaining themselves in a steady state can avoid the increase of entropy and, further, may even develop toward states of increased order (negentropy).[1,2] This is the basis of Schrodinger's statement that life 'feeds on negative entropy.'

The steady state or, as it is also called, dynamic equilibrium, shows remarkable regulatory characteristics. This becomes particularly evident in the principle of equifinality. In any closed system the final state is unequivocally determined by the initial conditions; if either the initial conditions or the nature of the process is altered, the final state will also be changed. In open systems, however, this is not the case. Hence, the same final state may be arrived at from quite different initial conditions and in different ways. This is what is called equifinality, the attainment of equivalent final states in a variety of possible ways.

2. Entropy in Statistical Thermodynamics and Communication Theory

The interpretation of entropy in classical thermodynamics is often confused with its usage in statistical thermodynamics (or mechanics) and in communication (or information) theory.[3] The meaning of entropy in these fields will be briefly reviewed.

Statistical thermodynamics is based upon a hybrid foundation in which the rigidity of mechanical laws is interwoven with the uncertainty specific to the notion of probability (113, p. 7). In the view of statistical thermodynamics, the degradation of energy is not irreversible. There exists a particular probability, albeit a very small one, that a pile of ashes may reorganize its energy to the extent that it becomes capable of heating a furnace. There exists the probability, too, that a corpse may resuscitate to lead a second life in exactly the reverse order of the first. If we have not yet witnessed such 'miracles', it is only because we have not been watching a sufficiently large number of corpses or piles of ashes (113, p. 7).

The primary foundation of statistical thermodynamics is Boltzmann's formula that the entropy of an isolated gas of N molecules is given by:

$$\text{Entropy} = S = \ln W,$$

where

$$W = \frac{N!}{N_1! N_2! \cdots N_s!}$$

and the N_i values $N_1 \ldots N_s$ represent the distribution of the gas molecules among the s possible states. $S = \ln W$ has been translated as 'entropy is equal to the thermodynamic probability.' This interpretation of entropy has been extended to apply not only to energy, but also to material structures; in nature there is a constant tendency for order (heterogeneity) to turn into disorder (homogeneity). Disorder, then, continuously increases, rather than the quantity of bound energy, and the universe tends toward chaos, which is, in the view of Georgescu-Roegen, a far more forbidding fate than the Heat Death, the final attainment of the temperature equilibrium in the Universe. Entropy has become redefined as a measure of the degree of disorder in a system. This measure, in turn, is defined in terms of the probability of the occurrence of a particular distribution of energy or matter. Thus, physical events are directed toward states of maximum probability, and physical laws, therefore, are essentially 'laws of disorder', the outcome of unordered statistical events.[4,5]

A second interpretation of entropy that is distinct from its classical usage has arisen in communication or information theory. This interpretation is loosely based upon the usage of entropy in statistical thermodynamics. In 1948 Wiener (356) introduced a specific definition of the 'amount of information' in a probability distribution. In order to illustrate the concept of information, let us take the example of a card randomly drawn from a full deck of cards. If one wishes to guess which card has been drawn there are fifty-two possibilities to consider; there are fifty-two question marks in one's mind. If one is told that the card is a face card (a jack, queen, or king of any suit) forty of those question marks vanish. If one is further told that the suit of the card is spades, only three question marks are left. The smaller the proportion of the initial question marks left after some information has been made available, the greater the amount of that information. The general principle is that the amount of information, $I(E)$, that the event E of probability p has occurred is measured by the formula

$$I(E) = F(p)$$

where F is a strictly decreasing function which for obvious reasons may be assumed the condition $F = 0$ for $p = 1$. Wiener employed the function $-\log p$. Thus,

$$I(E) = -\log p.$$

At about the same time Shannon (270), a telecommunications engineer, arrived at a measure of the capacity of a code system to transmit information. Shannon's conceptualization of the problem of measuring information was similar to that of Wiener: the indecision of an observer who would have to guess the nature of one randomly chosen element in a set. In the game of Twenty Questions, where one is supposed to find the identity of an object by receiving simple binary (yes or no) answers to questions, the amount of information conveyed in one answer is the elimination of one of two alternatives, such as animal or non-animal. With two questions it is possible to arrive at one of four possibilities; e.g. mammal – non-mammal, then flowering plant – non-flowering plant. With three questions it is possible to arrive at one out of eight alternatives, and so forth. Thus, the logarithm at the base 2 of the possible decisions can be used as a measure of information. This measure of information, as formulated by both Wiener and Shannon, is formally identical to that of the statistical thermodynamic concept of entropy, $S = \ln W$. Recognizing this, Shannon labelled his measure of quantity of information 'entropy', opening the door to the present confusion over what entropy 'means'.[6]

3. Entropy and Information: Summary

The three formulations of entropy, the classical, the statistical, and the informational, are quite distinct. There are, however, obvious links between them. In classical thermodynamics, entropy refers to the process of the irreversible degradation of energy in a closed system. By extension, it is also used to describe the degree to which that process has reached its final state of complete degradation. That is to say, entropy may serve as an index of the relative amount of bound (or, conversely, free) energy in a closed system. Since the quantity of energy in a closed system must be conserved, the amount of bound energy also implies the general nature of the distribution of all energy within the system. A large amount of bound energy (high entropy) implies a homogeneous distribution of energy; a small amount of bound energy (low entropy) implies a heterogeneous distribution of energy. The distribution of energy, in turn, signifies the relative degree of order of the *energy* of the closed system.

In statistical thermodynamics or statistical mechanics (a broader term that encompasses matter as well as energy considerations) the primary concept is the distribution of *energy or matter*. More precisely, it is concerned with quantifying the qualitative differences in the states of a system. Boltzmann's formula defines entropy as the probability of the energy or matter in a system attaining a particular configuration given a number of alternative configurations. The general tendency in nature is for matter and energy to go from less probable to more probable configurations, from order (heterogeneity) to disorder (homogeneity). The exception to this rule (and to the Second Law in closed systems) is the behavior of an open system, which can maintain itself in a state of high statistical improbability, that is, order. In the statistical interpretation, then, entropy is equated with systemic order.

In communications theory, the emphasis is on the information contained in a probability distribution. The measure of the amount of information is isomorphic to the measure used in statistical mechanics to identify the order in a system, and has been labelled entropy. Thus, 'entropy' or the information statistic has become a generalized measure of order, the degree to which the elements of any set distributed in real or conceptual space depart from randomness. This has major implications as to the predictability of systems. In a highly ordered system, if the value of one element is known, the value of all other elements should be able to be predicted. On the other hand, in a highly disorganized system even if the values of $n - 1$ elements are known the value of the nth is not predictable.

There is, then, a logical progression in the mutation (or is it mutilation?) of the entropy concept; from the process of energy degradation it has become the distribution of matter and energy and, finally, a measure of the configuration of any probability distribution. This has led many writers, including such diverse types as the geomorphologists Chorley and Kennedy (69, p. 219) and the physicist Brillouin (45, p. 159), to suggest that entropy and probability are synonymous. As we have seen, this may not be completely accurate.

4. Organization as Spatial Process
a. Order

For all practical purposes the terms organization and order may be regarded as equivalent. An organized system is an ordered one; a disorganized system is a disordered one. Not only is this equivalence implicit in the alternative formulations of the entropy concept that have just been reviewed, but numerous authors also explicitly make this point.

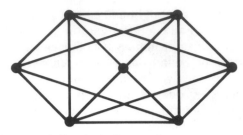

A. Disordered system (High entropy): Topologically complex.

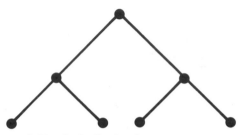

B. Ordered system (Low entropy): Topologically simple.

Figure 8.1 *Topological complexity of ordered and disordered systems*

Bertalanffy (28, p. 151), for example, states that the basis of organization is order, and Khailov (169, p. 30) notes the widespread tendency to regard organization as order and to evaluate it by a rigorous measure of order, negentropy. The 'widespread tendency' to which Khailov refers is probably a direct result of the abstract, even somewhat amorphous, nature of the concept of organization. Order is both a manifestation of organization and an objective measure of it. A system that has undergone an organizing process is spatially ordered (heterogeneous); one that has undergone a disorganizing process is spatially disordered (homogeneous).[7] The probability measures developed in statistical mechanics and refined in communication theory provide a method for quantifying order and, by extension, organization. In chapters 4 and 6, examples of the application of the information statistic were presented. Chapman (64), Marchand (204), Gould (127), and Medvedkov (209), to name just a few authors, provide more comprehensive illustrations of the measurement of 'entropy' within the context of geographical analysis.

Before continuing further, it is of more than passing interest to note that one other useful tool for the measurement of order or organization exists; this is topological analysis. According to Thom (307, pp. 135 ff.), entropy is related to the topological complexity of a system. The higher the entropy of a system, the greater is its topological complexity. Highly ordered (low entropy) systems, a hierarchically structured one for example, are topologically less complex than disordered systems (Figure 8.1). Order implies a stable system (173, p. 62). Topologically less complex systems are, therefore, more stable; they are better able to resist a perturbation of a given amplitude than are systems of more complex form. Further, the evolution of systems will be dominated by an overall tendency away from less stable forms, away from those less ordered systems in which there is total connectivity (188, p. 115; 307, p. 144). But this is anticipating, without proper foundations being prepared, the examination of hierarchic structure that occurs later in this chapter. In closing, one might point out that the appearance of topology in the present context substantiates in part the comments in the early portions of this book regarding the value of generality and abstraction.[8] More specifically, this utilization of topological analysis furnishes a great deal of support for Rashevsky's contention (see chapter 5) that topological relations are more fundamental to understanding nature than are those involving geometry.

b. Organization and Disorganization

The entropic tendency is perhaps the most fundamental natural law, affecting all organic and inorganic elements of the universe, and even specifying the direction of time. By extension, this original meaning has come to include an index of the distribution of energy or matter in a system; this is the concept of order. In the formulation of the organization process presented here, both of these facets of entropy are employed. That is to say, there is a concern with both the spatial distribution of energy and matter, order, and the means through which the elements of a set become either more or less ordered in space.

The fundamental influence upon the distribution of objects and events over the surface of the earth at both the macroscopic scale, involving the arrangement of discrete elements, and the microscopic scale, involving the change in the intrinsic morphology of one discrete element (this includes growth), is the entropic process of organization. The concepts of both organization and entropy are generic. Thus, objects and events may tend toward spatial order, a negentropic process, or toward spatial disorder, an entropic process. The absence of a tendency in either of these directions indicates that a stable equilibrium has been attained; being has replaced becoming. The direction which the process will tend to take is a function of the nature of the system, especially of its relation to its environment. In a closed system, the destruction of order is the general direction of events. In an open system, on the other hand, the direction may be reversed and order increased.

The sharp distinction between open and closed systems that is commonly employed is inadequate and misleading, particularly where human systems are involved. We may more properly consider many systems to have characteristics that are valvular or, resorting to mythology, Januan.[9] These systems open their boundaries to the import and export of matter and energy in certain situations and close them in others. Although the terms 'open' and 'closed' system will continue to be employed here, they are meant to imply the valve-like nature of many systems.

The difference in the direction of orderliness toward which open systems and closed systems move is in large measure a function of the nature of the primary feedback mechanism of each type. The principle of feedback is, of course, an old one, used by James Watt to keep the velocity of his steam engine steady under varying load conditions. In more recent times, the application of the feedback principle has fallen under the name of 'cybernetics', a derivation of the Greek *cybernitos* or helmsman. Perhaps the simplest illustration of the cybernetic principle is a thermostatically controlled heating system. When the temperature falls below a certain level the rate of production of heat in the furnace is increased; when the temperature rises above a certain level the production of heat is retarded.

Since its inception, cybernetics has been more or less identified with self-regulating or equilibrating systems. The very term 'helmsman' makes it quite explicit that the primary concern of the field is with *deviation-countering* systems such as the steam engine and furnace examples above. Less attention has been paid to systems in which the relationships between components are *deviation-amplifying*. The former class of systems is said to possess negative feedback characteristics, while the latter possesses positive feedback characteristics. Both of these broad classes of systems are mutual causal systems. That is, in these types of systems the elements influence each other either simultaneously or alternatingly (206, p. 6). The relative neglect of deviation-amplifying systems has led Maruyama (206) to write of 'the second cybernetics', in which the role of positive feedback is emphasized. The recognition of this 'second' cybernetic relationship is important since a given system will likely contain both positive and negative feedback loops. That is, one subsystem may be becoming more highly organized while a second tends toward its most probable state.

Those spatial systems that tend toward disorder are, then, closed systems predominantly characterized by negative feedback; those that tend toward order are open systems characterized by positive feedback. The direction of the feedback loops, in turn, depends upon the specific nature of the causal relationships between the components of the system (21, p. 160). It is extremely important to attempt to learn more concerning these small scale relations, as they form the basis of the behavior of systems at larger scales. The investigation of these micro-relations is, however, in the domain of a viewpoint that is quite distinct from that adopted in this discussion. To borrow an analogy from biology, such matters are like the study of the chemistry of proteins while the present approach is more similar to the study of an organism's behavior. This analogy may be continued to make one further point. The existence of positive and negative feedback relations suggests the existence of some form of control[10] or entelechy. This control is not due to an external force but is, rather, an intrinsic phenomenon that guides a system in a manner similar to that in which genes regulate the rate of cell growth of an organism. It can be strongly argued that this control is related to the concepts of optimality and equilibrium. The role of these concepts in organization will be examined in more detail below.

Organization and growth are closely related phenomena and feedback provides the direct link between them. Growth, as we saw in the preceding chapter, is a generic term which refers to positive or negative growth, or to the absence of growth. In any system which is undergoing positive growth or morphogenesis, deviation-amplifying mutual causal process predominates; energy and matter become more highly organized and assume a stable, topologically non-complex form, perhaps an hierarchical one. In negative growth or morphostasis, deviation-countering feedback predominates; energy and matter become less organized, assuming a complex structure that is relatively unstable. In either type of growth, however, both forms of feedback are likely to be present either simultaneously or alternating with one another. The less dominant form of feedback provides the necessary constraints upon the nature and duration of the more dominant. In other words, there is an interplay of *fixed rules* set by the constraints and the *flexible strategies* through which the final form can be attained (173, p. 62). Finally, in the case of a no-growth system, a stable equilibrium characterizes the system and there is a tendency toward neither further organization nor further disorganization.

Systems undergoing positive growth are generally open and tend toward spatial order; this implies that they are more stable. One way in which this stability manifests itself is in the existence of stable sub-assemblies such as the interlocking levels of a hierarchy. In all dynamically stable systems, steady state systems in other words, stability is maintained by the equilibrium of opposing forces: positive vs. negative feedback; centrifugal vs. centripetal forces. In Koestler's (173, p. 62) view, this is not the reflection of any metaphysical dualism but, rather, of Newton's Third Law: to every action there is an equal and opposite reaction.

The development of a city on an agricultural plain is an obvious example of the way in which deviation-amplification creates order in space. In the beginning, the plain is largely homogeneous in terms of both its distribution of population and its potentiality for agriculture. By some chance a farmer may open a tool shop which eventually becomes a meeting place for other farmers. A food stand might be established next to the tool shop and gradually a village grows. Since the village facilitates the marketing of agricultural products, additional farms may flourish in the immediate vicinity. The increased agricultural activity then necessitates the development of industry and services, and the village grows into a city. The internal order of the city increases, as does the order of the formerly homogeneous plain. Both attain highly improbable states.[11]

At a larger scale (a smaller map scale), however, a deviation-countering tendency may be promoting disorder in space. The growth of the city will likely have an inhibiting effect upon the growth of another city in the immediate vicinity. A city requires a hinterland to support it and cities are, therefore, generally spaced at some intervals. If several cities of similar size develop in this larger region, we may expect them to do so in such a way that they will be uniformly spaced with respect to one another, in order that each may establish a sufficiently extensive hinterland for support. The larger region thus becomes spatially disordered. Now there is order at one scale and disorder at the second. If this simple model is modified in such a way to allow cities to grow to different sizes and, therefore, to perform various ranges of functions and serve hinterlands of various sizes, we have arrived at the familiar central place arrangement or hierarchy.[12]

This is the second time that the concept of the hierarchy has been encountered. The first was in the context of the topological analysis or the relative degree of order in a spatial system, where it was noted that hierarchically arranged systems are more stable. As suggested at that time, the hierarchical arrangement is perhaps the most fundamental manifestation of positive organization and one of the most common structures found in nature. In view of this, the concept of hierarchy needs to be examined in more detail.

B. SPATIAL HIERARCHIES

Organized complexity frequently manifests itself in the form of a hierarchically ordered structure. Numerous authors in the physical, social, and biological sciences and in the various branches of philosophy have noted the ubiquity of hierarchical organization among both natural and artificial phenomena. Smith (278, p. 81) even goes so far as to state that 'any aggregate that is neither completely ordered nor completely disordered must have hierarchical aspects.'

Both phenomena in space, for example a city or a fluvial system, and phenomena that are generally regarded as aspatial, such as the 'chain of command' or 'pecking order' in various types of social organizations, may manifest themselves in hierarchical order. One commonly applied distinction between spatial and non-spatial hierarchies is that of structural versus functional hierarchies. This is misleading for two reasons. First, as we have seen, spatial order implies flow or movement as well as structure; flows, too, may be hierarchically ordered. And, second, structure and function represent complementary aspects of an indivisible continuum. This question of the relationship between structure and function is, of course, one of the classical issues in both philosophy and biology. Finally, and not least in importance, the concept of hierarchical arrangement transcends the boundaries between disciplines in that it focuses upon the general interrelationships of structure, movement and function in a wide variety of phenomena, rather than isolating the specific characteristics of a particular phenomenon. In other words, hierarchies have common properties that are independent of their specific content. In this sense, the concept is especially relevant to the theme and goals of the present discussion. A more detailed examination of spatial hierarchies commences with a brief review of some of these common properties.

1. What Are Hierarchies?

One of the clearest illustrations of the essence of hierarchical arrangement is Simon's (274, p. 5) example of a set of Chinese boxes of a particular

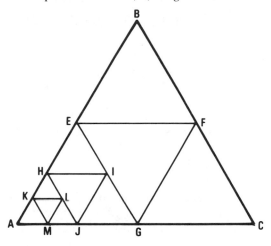

Figure 8.2 *Hierarchic structure*

kind. A set of Chinese boxes usually consists of a box enclosing a second box which, in turn, encloses a third; the recursion continues as long as the patience of the craftsman holds out. Hierarchies are a variant on this pattern. Opening any given box in a hierarchy discloses not just one new box within, but a whole smaller set of boxes. Each of these component boxes, in turn, contains a whole new set. While the ordinary set of Chinese boxes is a sequence, or complete ordering, of the component boxes, a hierarchy is a partial ordering and may be likened to the branching of a tree. Figure 8.2 illustrates the notion of successively more encompassing sets.

In a hierarchy, a given set or level must be described not only *per se*, but also in terms of both what is subsumed by it and what subsumes it. In Figure 8.2, triangle AEG consists of four sub-triangles, one of which is AHJ, and it, in turn, consists of four sub-triangles. On the other hand, AEG is but one of four sub-triangles that compose ABC. A hierarchical arrangement is, then, characterized by relativity where the terms 'part' and 'whole' are concerned. Wholes and parts in an absolute sense just do not exist anywhere (173, p. 48). What we find instead are intermediary structures on a series of levels in an ascending order of complexities: sub-wholes which display, according to the way that you look at them, some of the characteristics commonly attributed to parts and some of the characteristics commonly attributed to wholes. Noting that the elements in a hierarchy, like the Roman god Janus, all have two faces looking in opposite directions, one toward the lower level 'parts' and one toward the higher level 'wholes', Koestler (173, p. 48) identifies this 'Janus effect' as a fundamental property of all types of hierarchy. In addition, he coins a new word to describe the Janus-faced entities: holons, from the Greek *holos* (whole), with the suffix -on which, as in proton or neutron, suggests a particle or part. In geography, Berry's (21) conceptualization of 'cities as systems within systems of cities' is an explicit recognition of the Janus effect.

The nesting of subsystems within larger subsystems that has been illustrated is related to a second fundamental property of hierarchic order. Simon (274, p. 9) refers to the second property as 'near decomposability'.[13] Briefly, this concept refers to the fact that the interactions that occur in nature, between and within systems of all kinds, decrease in strength with distance. Hence, any given component has most of its strong interactions with components that are in proximity (in either real or conceptual space), many of them at the same level. As a result, a system is likely to behave either as made up of a collection of localized subsystems or as a more or less uniform entity with relatively strong internal cohesion. Systems of the former kind are hierarchical in nature. Returning to the 'cities as systems within systems of cities' notion, it is possible to analyze a city or a regional urban system outside of the broader context of the national or global urban system of which it is an element because it is 'nearly decomposable'.

Having considered these general properties of hierarchical systems, we next focus upon spatial hierarchies and their specific manifestations.

2. Hierarchical Organization of Spatial Systems

A useful framework within which to examine spatial hierarchies is one employing the concept of dimension (see chapters 4 and 5). On this basis, it is possible to distinguish between hierarchies of points, of lines, and of areas.

a. Hierarchies of Points

Figure 4.23A depicts a hierarchy of points which are distributed over a surface, cities of differing size on a plain. Since points have a dimension in length of zero, however, it is not strictly correct to refer to this as a spatial hierarchy of points. Here population size or number is the distinguishing feature. The points themselves, however, have no differentiable spatial characteristics; size categories are the primary criteria rather than spatial relations. In order to deal with a hierarchy of point size one must revert to an aspatial formulation, the rank–size rule.

The rank–size rule can be stated as

$$P_n = P_1 (n)^{-1}$$

where P_n is the population of the nth city in the series 1, 2, 3, . . . n in which all towns in a region are arranged by population in descending order, and P_1 is the population of the largest city. The population of the nth ranking city is equal to the population of the largest city divided by the rank n. For example, the population of the tenth ranking city in an urban system whose primate city has a population of ten million should be

$$10,000,000 \times 10^{-1} = 1,000,000.$$

Since this relationship graphs as a straight line on double-logarithmic graph paper, it is a member of the class of Pareto distributions. Work by Berry

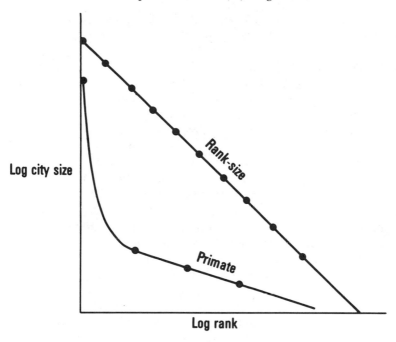

Figure 8.3 *Rank–size and primate distributions of urban population*

(22) suggests that Pareto urban population distributions are characteristic of countries which are: (a) larger than average, (b) have a long history of urbanization, and (c) are economically and politically complex. Countries that do not posses these attributes are often characterized by a primate distribution (Figure 8.3).

Pareto distributions are exceedingly common and apply not only to human settlements but also to the numerical distributions of plant species, dog breeds, word frequency in human speech, and diverse other phenomena. This has led Simon (273) to suggest that the rank–size regularity is the manifestation of a set of stochastic relationships that affect the development of any phenomenon. More specifically, Simon believes the rank–size distribution to represent the equilibrium slope of any general growth process in which each element initially has a random size and thereafter grows in an exponential manner that is proportional to its size. Moreover, the stability of the rank–size relationships in various systems over both time and space suggests that it may be the manifestation of a steady–state condition in which the distribution is affected by a myriad of small random forces.

In chapter 7 the rank–size rule was examined in the context of the constant ratio aspect of the allometric growth of a system. Since it has been established that the allometric principle is an expression of system optimality, it is not unreasonable to suggest that the rank–size distribution may be similarly related to some kind of optimization principle in social-economic systems.

Now that a non-spatial hierarchy of points has been examined, let us consider the case where the distinguishing characteristic of a set of points distributed in space involves not some aspatial magnitude but, rather, intrinsically spatial properties. Since, as has been noted, points have the dimension in length of zero, these spatial properties will necessarily have to be topological (non-metrical) rather than geometrical. And since the primary attribute of topology is connectivity, a hierarchic arrangement of a set of points will be revealed through the connectivities between points. These connectivities may be manifest either in physical links, such as roads, or in the flow of less tangible phenomena, such as ideas or information, in its general sense. Figure 8.4 illustrates hierarchic and non-hierarchic point sets. The distinction between Figures B and C may be conceptualized in terms of the depth and span of hierarchies. The branching hierarchy has, in this case, a depth of three levels and a span of six elements, while the direct control hierarchy has a depth of two levels and a span of fourteen elements. In non-spatial terms, the former may be said to be representative of a complex economy in which a rank–size urban population distribution would be expected, and the latter representative of a less complex economy dominated by a primate city.

A better appreciation for the properties of a point hierarchy can be gained through an analysis of the centrality or accessibility of each point. Figure 8.5 illustrates the connectivity matrices for a branching hierarchy and for a non-hierarchical point set or 'chain', each of seven elements. The row sums at the right of each matrix indicate the degree of each point, the total number of *direct* linkages from a given point to other points. This is a primitive measure of centrality, yet it is obvious (as it is solely on the basis of visual inspection) that certain points in the hierarchical arrangement are more central than any of the points in the chain. In this instance, the middle level points (2 and 3) are most central.

The accessibility matrices for these two point sets are shown in Figure 8.6. An accessibility matrix is determined by summing the connectivity matrices taken to the powers 1 to n, where n equals the diameter of the network, the number of edges in the shortest path between the most distant points.[14] Such an accessibility matrix enumerates the total of all *direct and indirect* connections in the point set. The row sums indicate the relative accessibility of each node to all other nodes in the network. The higher the value of the row sum, the greater the accessibility of the node. Figure 8.6 shows that in the hierarchy point 1 has the greatest centrality; it is in the optimum location from which to interact with all other points. The degree of centrality of the 'apex' point relative to points on the chain set can be expected to increase in hierarchies that have greater depth. The centrality of the points on the chain set can be seen to symmetrically decline from point 4, the middle point. The row sums of the two matrices

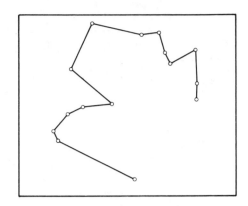

A. Non-hierarchic or "chain" point set.

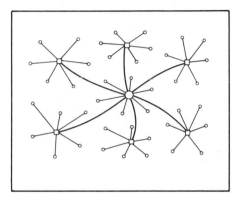

B. Branching hierarchy.

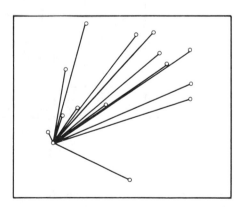

C. Direct control hierarchy.

Figure 8.4 *Hierarchic and non-hierarchic point sets*
    Source: Abler, Adams, and Gould (2, pp. 260–4)

	1	2	3	4	5	6	7	Σ
1	0	1	1	0	0	0	0	2
2	1	0	0	1	1	0	0	3
3	1	0	0	0	0	1	1	3
4	0	1	0	0	0	0	0	1
5	0	1	0	0	0	0	0	1
6	0	0	1	0	0	0	0	1
7	0	0	1	0	0	0	0	1

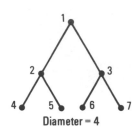

A. Hierarchical point set.

	1	2	3	4	5	6	7	Σ
1	0	1	0	0	0	0	0	1
2	1	0	1	0	0	0	0	2
3	0	1	0	1	0	0	0	2
4	0	0	1	0	1	0	0	2
5	0	0	0	1	0	1	0	2
6	0	0	0	0	1	0	1	2
7	0	0	0	0	0	1	0	1

B. Non-hierarchical point set.

Figure 8.5 *Connectivity matrices of hierarchical and non-hierarchical point sets*

are not comparable, as matrix *A* was powered only four times and matrix *B* six times. This example clearly illustrates the essence and implications of a spatial hierarchy of points.

b. Hierarchies of Lines

i. Methods of Ordering Linear Hierarchical Systems. The hierarchical manifestation of a set of lines may be regarded as a topological tree, much like that used to illustrate a hierarchy of points (Figure 8.5A). The primary distinction between a branching hierarchy of points and a branching hierarchy of lines (Figure 8.7) is that in the former the points represent the

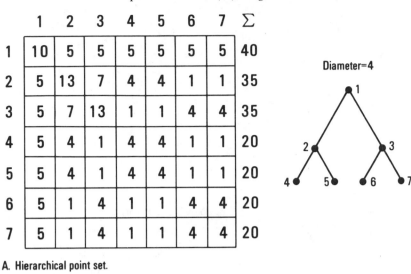

	1	2	3	4	5	6	7	Σ
1	10	5	5	5	5	5	5	40
2	5	13	7	4	4	1	1	35
3	5	7	13	1	1	4	4	35
4	5	4	1	4	4	1	1	20
5	5	4	1	4	4	1	1	20
6	5	1	4	1	1	4	4	20
7	5	1	4	1	1	4	4	20

A. Hierarchical point set.

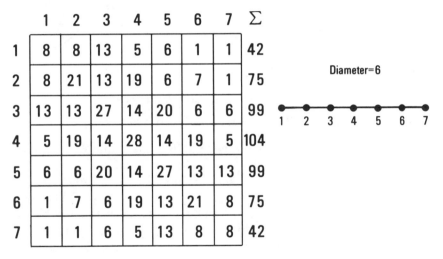

	1	2	3	4	5	6	7	Σ
1	8	8	13	5	6	1	1	42
2	8	21	13	19	6	7	1	75
3	13	13	27	14	20	6	6	99
4	5	19	14	28	14	19	5	104
5	6	6	20	14	27	13	13	99
6	1	7	6	19	13	21	8	75
7	1	1	6	5	13	8	8	42

B. Non-hierarchical point set.

Figure 8.6 *Accessibility matrices of hierarchical and non-hierarchical point sets*

location of some set of 'concrete' elements; in the latter case points are implicit and of secondary concern, denoting the end points or junction of lines (bounding the lines, in other words). Although, by definition, the edges of a topological tree are without magnitude, it is clear that the edges of real world spatial systems, say a river or a railway system, have differing magnitudes and therefore require a method by which they may be distinguished (365, p. 3).

Perhaps the most commonly employed example of a linear spatial system is a fluvial system. In order to draw upon some well-developed

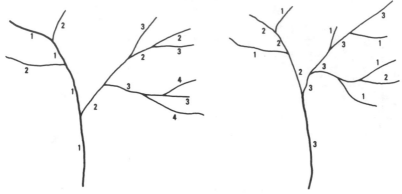

A. Gravelius' stream orders.

B. Horton's stream orders.

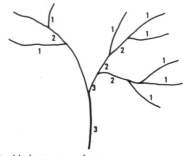

C. Strahler's stream orders.

Figure 8.7 *Methods of ordering stream segments:* R_b = 3 *stream*

illustrations from the geological and geographical literature, rivers will be employed in the present examination of linear hierarchies.

Gravelius (130) was among the first of the modern scholars to attempt to assign order numbers to the stream segments of a fluvial system. The main trunk was designated order one and smaller tributaries were given higher order numbers (Figure 8.7A). In this method the identity of stream one is maintained through a junction along the branch whence comes the greatest flow. The other branch is stream two. The problem with this ordering method is that the order number is not related to the magnitude of the segments. Stream segments which connect end points to junctions may be of any order.

Gravelius' primitive ordering system was subsequently modified by Horton (154), who assigned to the main stream the highest order and the unbranched fingertip tributaries the order one (Figure 8.7B). A second order stream is formed by the junction of two first order streams, and may be joined by other first order streams. A third order stream is formed by the junction of two second order streams and may be further joined by other first and second order tributaries. More generally, the junction of two streams of like order forms a stream of the next highest order, which can receive tributaries of any order lower than its own. Through this method order becomes a function of the magnitude and number of streams. It is possible, however, to have instances where streams of similar physical magnitude have orders which are different.

Several years after Horton's formulation, Strahler (298) proposed a further modification of the method of assigning order numbers to streams. Strahler's system involved assigning orders to stream segments in a fashion similar to Horton's preliminary step, but these were the final orders to be considered by Strahler (365, p. 7). In this way, stream segments of similar magnitude are assigned the same order and are unambiguously numbered (Figure 8.7C). More recently, Shreve (272), Scheidegger (266), and Woldenberg (365) have suggested further refinements of Strahler's system. These have been attempts to correct the major shortcoming of the Strahler system – the lack of its ability to distinguish between a stream of order U which is fed by two $U - 1$ order streams, and a stream of order U which is fed by more than two lower order streams. The alternative methods proposed generally employ ratios of the logarithm of order number to the logarithm of the bifurcation ratio (see below). These will not be described here, as they do not differ significantly from the basic method, which has the added advantage of relating each segment of order U to a drainage basin of order U. The reader is referred to Woldenberg (365, chapter 1) for a review of these more sophisticated systems.

This general approach to ordering linear hierarchies involves only the topology (what is connected to what) and geometry (magnitude defined in terms of relative length) of a branching system. As such, it has very broad utility and has been successfully applied to such diverse spatial phenomena as Alpine glaciers and organic trees (365), and pulmonary and arterial systems (370). In addition, one could apply it to a wide number of additional spatial systems, an urban road network where the dominance of one edge over another (its magnitude) is signified by a stop sign.

ii. The Bifurcation Ratio. In chapter 7, the concept of a bifurcation ratio was introduced in the context of the inverse geometric series which characterizes allometric growth. More specifically, the bifurcation ratio indicates the particular form of the geometric progression. A bifurcation ratio of three, for example, indicates that for each order below the highest the number of elements having a particular order number is a threefold increase over the preceding number. In this instance, the number of streams N_u having any order u is described by

$$N_u = 3^{(k-u)}$$

where k is the highest order and 3 is the bifurcation ratio. (See chapter 7 for an illustration of this principle.) In the present context of hierarchical order, the bifurcation ratio may be conceptualized as an index of the span of the hierarchy, the total number of elements at each level. Thus, with a bifurcation ratio of three, the first level below the highest may be expected to have $3^1 = 3$ elements; the second below the highest, $3^2 = 9$ elements; the third below the highest, $3^3 = 27$ elements, and so forth.

iii. Branching Angles. A corollary to the preceding discussion of the ordering and span of branching hierarchies concerns the angle at which these linear systems branch. Lubowe (193) has demonstrated that in the case of rivers flowing over lithologically and structurally homogeneous surfaces in different areas of the USA, 'Mean junction angles in each area increase as the order of the receiving stream increases.' Thus, the angle made when a first order stream flows into a third order stream will be greater than the angle formed when a first order stream flows into a second order stream. Moreover, the angle will vary directly with the order of the receiving stream for a given order tributary. Using an elm limb as his case study, Howie (156) found a similar general regularity in branching angles.

The formulation of this principle is really no more than a logical extension of the early (*c.* 1925) work of Cecil Murray (214), who demonstrated that in blood vessel systems the smaller the arterial branch the closer to ninety degrees will be its angle of divergence from the trunk. Murray further demonstrated that this regularity is a result of the principle of 'minimal work' (least effort), which he deemed the 'very criterion of organisation.'

c. Hierarchies of Areas

In the above discussion of hierarchies of points, it was established that the delineation of spatial hierarchies must be based solely upon the spatial (geometrical, topological, or dimensional) properties of a system. The rank–size rule was then demonstrated to be an inherently aspatial hierarchy, since it is based directly upon population size. Extending this reasoning, it may be strongly argued that, in the terms in which it is conventionally formulated, the central place hierarchy of cities is inherently aspatial. The specification of a hierarchy of cities is a direct function of a non-spatial property, the size of each individual city in terms of either its population or the number of functions that it provides. Without the inclusion of city size the central place 'hierarchy' would be no more than an undifferentiated web (rather than a branching system which, as we have seen, is differentiable in terms of hierarchic levels) of points.

In order for the central place hierarchy to have a spatial conceptualization, it must be formulated in terms of a hierarchy of market areas, rather than of cities.[15] Market areas may be differentiated on the basis of the geometrical attribute of area (L^2); higher order market areas are larger than lower order market areas, and market areas of the same order have equal sizes. The size of each market area is, of course, proportional to the population size of the center that it supports. Although Woldenberg (365, p. 69), for example, explicitly conceptualizes central place theory in the conventional way, in terms of a 'hierarchy of cities' (points), it is implicit in his analysis that the hierarchy actually involves the market areas of the central places. Thus, a central place hierarchy manifests itself in a successive nesting of hexagonal market areas.[16] The areal extent of each hexagon at a particular level of the hierarchy is, of course, directly related to the threshold of demand required to support the range of functions available at the central place which serves it.

In the hierarchy of hexagonal areas, there exists a relationship that is similar to that described by the bifurcation ratio of stream segments. This is known as the ratio of areas, R_A, and is generally stated as representing the ratio of the area of a hexagonal market area to the area of one of the next lower order hexagons nesting within it. Figure 8.8 illustrates three of the most commonly found ratios in a central place system. Note that the variation in the ratio is accomplished by varying the size and orientation of the higher order hexagon.

The ratio of areas may logically be interpreted in an alternative manner: R_A expresses the total number of lower order hexagons (the total is arrived at by summing hexagons and portions of hexagons) within the higher order hexagon.[17] This interpretation of R_A is isomorphic to the bifurcation ratio in that it expresses the number of $u - 1$ order elements that we can expect to be associated with each u order element. In the case of the market areas, R_A is also an equivalent way of expressing the K value of a central place hierarchy. Recall that the K value is conventionally conceptualized as the total number of settlements served by each central place, rather than the number of market areas of the next lower order, as is proposed here. Therefore, the configurations illustrated in Figure 8.8 correspond to: (a) the marketing principle ($K = 3$), (b) the transportation principle ($K = 4$), and (c) the administrative principle ($K = 7$). Further, as in the case of the bifurcation ratio, R_A may be used to derive the span of the hierarchy, the total number of elements at each level.

Since a complex social system would need to economize on market centers ($K = 3$ principle), transportation routes ($K = 4$ principle), and administrative linkages ($K = 7$ principle), it is not unreasonable to expect that all three principles operate simultaneously in geographic space. Some years after his original formulation of central place theory, Christaller came to recognize this and suggested a mixed hierarchy based on an R_A of 3.3. Woldenberg (365; 368) developed this notion further, utilizing a convergent mean of arithmetic and geometric means to derive a general theory of mixed hexagonal hierarchies. This model of mixed hexagonal hierarchies makes it possible to predict the numbers of areas of each order in an actual central place system. Woldenberg's approach will be considered in more detail below.

3. Why Are Hierarchies?

Now that the general nature of a hierarchy has been examined, and the hierarchical organization of spatial systems has been considered, the logical question to treat next is that of why hierarchies develop and persist. Two interrelated approaches to this question will be presented. The first represents the view of systems theorists concerning hierarchies in general; the second presents the more specific view of spatial systems theorists.

a. The Systems View

Simon (274) demonstrates on quite simple and general grounds that the time required for a complex system containing K elementary components

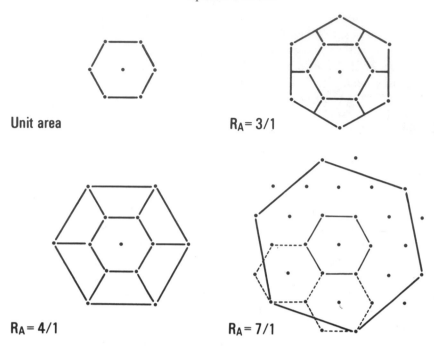

Unit area　　　　　　　　$R_A = 3/1$

$R_A = 4/1$　　　　　　　　$R_A = 7/1$

Figure 8.8 *Three possible hexagonal hierarchies*

to evolve by processes of natural selection from those components is very much shorter if the system is composed of one or more layers of stable component subsystems than if its elementary parts are its only stable components. The mathematics of the matter is a straightforward exercise in probabilities, but Simon illustrates the essence of his argument even more simply by a 'parable'.

Two watchmakers who assemble fine watches containing ten thousand parts each are frequently interrupted to answer the telephone. The first watchmaker has organized his assembly operation into a sequence of sub-assemblies. The average interval between telephone interruptions is a time long enough to assemble about 150 elements. An interruption causes any set of elements to fall apart completely. By the time that he has answered about eleven phone calls, the first watchmaker will usually have finished assembling a watch. The second watchmaker, however, will almost never succeed in assembling one; he will 'suffer the fate of Sisyphus.'

It has been argued on information–theoretic grounds that organisms are highly improbable arrangements of matter; so improbable, in fact, that there has hardly been time enough since the earth's creation for them to evolve. Simon points out, however, that the calculation on which this argument is based does not take into account the hierarchic arrangement of stable sub-assemblies in the organisms that have actually evolved. 'It has erroneously used the analogy of the second, unsuccessful watchmaker.'

When the first watchmaker is substituted for him, the times required are reduced to much more plausible magnitudes. Thus, hierarchies will evolve much more rapidly from elementary constituents than will non-hierarchic systems containing the same number of elements.[18] For this reason, almost all large and complex systems tend to have a hierarchic structure.

Complementary views on the formation of hierarchies are proposed by several other authors. Grobstein (133) suggests that hierarchical levels are generated by the variations in the collective interactions between the most elementary components of a system. The transition from a more elementary level to a higher level he calls a change of context. The structure generated by one set of elements and their interactions become reinterpreted or re-read in the context of a larger set of interactions to form a higher level. That is, the elements at any level of the system are not limited by their inherent and interactional properties at that level, but may generate emergent properties in the new context of a larger set. This has obvious analogies with the 'scale problem' in geography.

Levins (188) argues that very complex systems are dominated by processes that are self-simplifying and, in fact, tend to persist only in simplified modes of behavior. Hierarchical organizations are not generated from simple components that evolve to greater and greater complexity, as Simon suggests. Instead, Levins conceptualizes a hierarchical structure as a self-simplification of an initially complex and chaotic system.

And, finally, Pattee (231) explores the notion that new levels of a hierarchy develop not through a continuous evolutionary process but, rather, spontaneously as the result of structural instabilities in a system. In simple mechanical systems, the concept of stability can be clearly defined in terms of the system's ability to return to its original trajectory after being perturbed. In dynamical systems, however, the behavior of a system beyond its stable regime is not describable, let alone predictable (231, p. 146); discrete and spontaneous shifts to new levels may occur. Note the relationship of this formulation to Thom's theory of catastrophes (see chapter 5), which is, in turn, a general mathematical theory of the structural stability of topological structures.

A general theme that may be identified in the above perspectives is that of evolution. That is, with the exception of Pattee, these authors view the development of a system in terms of a continuous change toward those forms that are least complex and, therefore, most stable – hierarchies. This emphasis on evolution is not only a function of the concern of these authors at the research level with organic systems, but also reflects the fundamental fact that organization is an evolutionary process. The organization of both organic and inorganic systems may be viewed as evolutionary. Any hierarchical structure represents both a record of its development and a framework within which future changes occur.

b. Woldenberg: The Spatial Perspective

Any consideration of the 'why' of spatial hierarchies must begin by emphasizing one point: spatial hierarchies are not simply structural forms based upon geometrical and topological properties but are, rather, dynamic phenomena the structural properties of which are both causes and effects of the movements which occur within and between system elements. That is to say, a spatial hierarchy is a hierarchy of structure and flow; it is spatial process. Perhaps no other example so clearly captures the essence of the general spatial systems framework. Note, further, the close correspondence between this formulation and process metaphysics, which holds that the universe is composed not of 'things' but of a complex set of smaller and larger flow patterns that are self-maintaining (238).

Perhaps the major contribution toward understanding the existence of hierarchical spatial systems has been Michael Woldenberg's investigation of linear branching and central place systems.[19] Recall that the bifurcation ratio (R_b) of a fluvial system and the ratio of areas (R_A) of a nested system of market areas are isomorphic, indicating the structural and functional homologies between these classes of systems. Utilizing data obtained from various fluvial, glacial, arterial, and pulmonary systems, and from central place systems in various nations, Woldenberg demonstrates that a 'mixed hierarchy' based upon a convergent mean of the geometric progression of elements derived from R_b values (R_A values in central place systems) equal to 3, 4, and 7 is the structural arrangement that reflects maximum probability of matter and energy and, therefore, the maximum entropy possible in an open system.[20]

At the same time, the mixed hierarchy also reflects considerations of least work. In order to arrive at a least work configuration of a flow system, a river or a human system in which consumers interact with suppliers, for example, an equilibrium must be attained between the minimization of diffusion costs (in the sense of the frictional losses due to overland movement across a surface) and the maximization of economies of scale. There can be no hierarchical flow systems without some degree of scale economies, since this allows the integration of inefficient energy gathering and distributing surfaces into one system of higher organization. If a river had only one large basin instead of a hierarchy of them, transportation of debris would be held up by the large frictional losses (365, p. 116). Further, if there were very small economies of scale possible, it would be possible to have a very large number of small flows, thereby reducing frictional losses. On the other hand, if there were very small frictional losses resulting from overland transportation, it would be possible to have several large, highly economical flows. These conditions are generally not possible, however.

Many small flows, then, minimize diffusion costs, but also minimize the benefits of scale economies; small flows are inefficient. A few large flows maximize scale economies, but also maximize diffusion costs. Woldenberg demonstrates that the former alternative is a configuration which characterizes an arithmetic mean hierarchy, and the latter alternative characterizes a geometric mean hierarchy. The convergent mean mixed hierarchy thus represents an equilibrium state of the competition between the tendency toward scale economies and the tendency toward minimum overland costs. Therefore, the mixed hierarchy represents the least work solution, and will possess the appropriate number of levels so as to minimize the physical quantity of work.

The mixed hierarchy is, then, the configuration of the elements of an open system which not only reflects maximum possible entropy but also represents a least work condition. Leopold and Langbein (184, p. 30) have independently affirmed this equivalence of least work and maximum entropy, noting that the mathematical form which describes such a system is a geometric progression. This is precisely what R_A and R_b represent.

Before concluding this discussion, the distinction between a mixed hierarchy and an unmixed hierarchy needs clarification. As we have seen, the mixed or convergent mean hierarchy is derived from the simultaneous consideration of the three most common values of R_A and R_b. In central place terminology these represent, respectively, the marketing, transportation, and administrative principles. It is not entirely surprising that nature or human behavior often departs from the simple theoretical formulation produced by a scientist. Thus, the mixed hierarchical model is a refinement, a better method for describing and predicting the real world manifestation of spatial systems; it is an improvement upon the simpler models of 'unmixed' hierarchies involving and R_A or R_b value of either 3, 4 or 7. Even these unmixed hierarchies are optimal spatial systems. The mixed hierarchy is simply an *optimum optimorum.*

In general, then, a hierarchical spatial system is a dynamic equilibrium state or, if you prefer, a compromise between the opposing tendencies of organization and disorganization, between complete agglomeration and complete disagglomeration (Figure 8.9). Since the vast majority of terrestrial spatial systems are open (or at least valvular) systems in which order is maintained or even increased, complete disorder is not a member of the solution set. In addition, as noted in chapter 6, in order for human or physical flows to occur a gradient must exist in the system. This gradient is, in part, a function of the existence of a hierarchy of system elements. On the other hand, complete order is rarely a member of the solution set either, as the maintenance or growth of a system is a function of its ability to compete successfully for the resources available in its environment. Under a condition of complete order the system is too concentrated to support all elements. And, from a functional viewpoint, complete order is indistinguishable from complete disorder in that a gradient is absent.

C. CONCLUSION

As previously noted, organization and growth are highly interrelated; the interrelationship is such that each concept provides to the other some degree of the explanatory detail which is missing when considered independently. Having examined these two major aspects of change in the spatial characteristics of any system, and having identified them as components of the proposed focus of the general spatial systems framework – spatial process – the discussion of the latter two concepts is

A. Agglomeration.

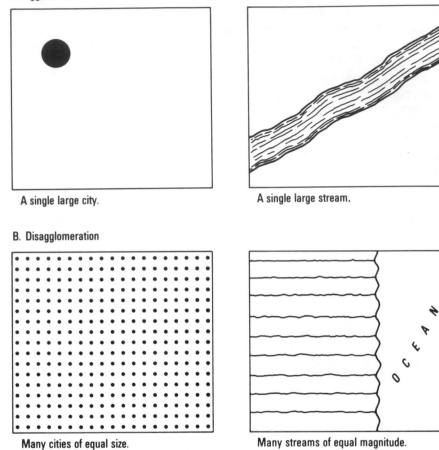

A single large city.

A single large stream.

B. Disagglomeration

Many cities of equal size.

Many streams of equal magnitude.

Figure 8.9 *Agglomeration and disagglomeration in human and physical systems*

concluded with some general comments concerning the role of process in geography. This occupies the first portion of the following chapter. The remaining portion of the chapter is devoted to an evaluation of the systems framework in geography.

Notes

[1] In a closed system, entropy always increases according to the Clausius equation: $ds \geq 0$. In an open system, in contrast, the total change of entropy can be represented by Prigogine's (243) equation:

$$dS = d_e S + d_i S$$

where $d_e S$ denotes the change of entropy by import and $d_i S$ the production of entropy due to irreversible processes in the system. According to the Second Law, $d_i S$ is always positive; $d_e S$ may be positive or negative.

[2] A change in the form of an open system may result; it may grow or decline if the flows which import or export energy between the system components, and

between the system and the environment, are imbalanced. A change may also occur if there is a modification in the relations of systems components. This is allometry (see chapter 7).

[3] For a lucid and more detailed discussion of the distinctions between the various interpretations of entropy, and of the confusion over their utilization, see Marchand (204).

[4] A cogent argument against this view that physical laws are laws of disorder is presented by Koestler (173); see especially chapter 10.

[5] This highly simplified account of statistical thermodynamics includes the essential points that are relevant in the present discussion. For a more detailed account, including a discussion of Maxwell's demon, see Georgescu-Roegen (113), especially chapter 6.

[6] A clear explanation of the probabilistic foundation of entropy is presented by Gould (127).

[7] This usage of spatial order and disorder is the converse of the intuitive connotation of the terms, in which one would expect an evenly spaced series of points to represent 'order' and a less regular or heterogeneous distribution to represent disorder.

[8] Recall that topology is the most general form of geometry, involving qualitative rather than quantitative relationships between the elements of a set. For further explanation, refer back to chapter 5.

[9] This reference to Janus alludes to his role in Roman mythology as the Gate Keeper, and is distinct from Koestler's (173, p. 48) 'Janus effect' which emphasizes the dual nature of 'holons'.

[10] A useful discussion of control systems and the role of feedback in the operation of these systems is contained in Bennett and Chorley (19).

[11] A similar example is employed by Maruyama (205).

[12] Gravity and potential models of relatively permanent human interactions, migrations in other words, are a similar manifestation of deviation amplifying feedback. The movement of individuals up the gradients on income and population potential surfaces increases the heterogeneity of the social–economic system.

[13] Koestler (173, p. 52) refers to this same property as 'dissectibility' and Pattee (231, p. 132) as the existence of 'discrete but interacting levels.'

[14] See Taaffe and Gauthier (301, chapter 5) for a more detailed explanation.

[15] As in the case of the elements of all cellular hierarchical systems, each market area is constrained toward a minimum surface of interface with other market areas. The figure of minimum interface per unit area which completely covers a plain is, of course, the hexagon. For a more detailed discussion of the significance of the hexagonal shape in human and physical spatial systems, see Woldenberg's paper, 'The Hexagon as a Spatial Average' (364).

[16] Since physical space may well be non-uniform because of differences in topography, transportation routes, and so forth, hexagonal market areas are rarely found on a conventional map of real space. Market areas behave as though they were hexagonal in transformed social space, however. See Warntz (335), Tobler (314), or Getis (115) for a discussion of this notion.

[17] In Figure 8.2, $R_A = 4$ signifies that: (a) each triangle of order u has an area four times larger than each triangle of order $u - 1$; and (b) each triangle of order u contains four triangles of order $u - 1$.

[18] The mention of time does not invalidate the concept of spatial process. As explicitly stated in chapter 1, all objects and events occur in both space and time. In some instances, however, it is convenient and more parsimonious to omit the temporal dimension and to focus on changes in the spatial relations of phenomena. Here, the intrinsic inseparability of time and space is simply being reaffirmed.

[19] Woldenberg's work appears in a number of important papers published in the *Harvard Papers in Theoretical Geography* series. See, for example (364; 365; 366; 367; 368; 370).

[20] The convergent mean is the mean of the geometric and arithmetic means of a set of data. In this case, each set of data is composed of three values corresponding to elements in geometric progressions to the base 3, 4, and 7, i.e. 3^n, 4^n, 7^n, where n is an integer power. See chapter 3 of Woldenberg (365) for a detailed explanation.

PART FOUR

CONCLUSION

9

Conclusion

In bringing this rather wide-ranging discussion to a conclusion an attempt is made to evaluate several of the major themes that have been touched upon in the preceding eight chapters. First, the notion of spatial process is considered; second, the range of systems approaches, both explicit and implicit, in geography; and, third, more specifically, the role of, and prospects for, the general spatial systems approach within the discipline.

I. Spatial Process

As Whitehead (353) has told us, the whole universe is characterized by an ongoingness which creates new entities out of a myriad of individual happenings. Although the disaggregation of this ongoingness into spatial process and temporal process may seem somewhat reductionist and, therefore, contrary to the general theme of this book, it has been argued that it is justified in that it is both parsimonious and a necessary step in attempting to discover the underlying principles of spatial order in the complex real world. Again, the term spatial process is not a denial that the universe, and most particularly that portion of the universe that is our home, changes both over time and space. Rather, the term suggests that a degree of understanding of aspects of reality may result from an analysis of the spatial properties of phenomena.

Spatial process may yield any of a set of possible configurations or structures. Each of these structures is but a transitory phase in Becoming; it is an entity on its way from one location in time and space to another location; it is growing or declining, organizing or disorganizing. The particular structure that we observe at a specific time and place is a function of the previous nature and location of the structure, and of the functional activities of the systems in which it is an element. The form of any spatial system is, then, both a record of its past and a framework constraining its future. In this sense, spatial process is an evolutionary process.

In the relatively brief consideration of spatial process presented here, several general and abstract concepts have been employed: geometry, topology, dimensionality, circulation, flow, diffusion, growth, organization, hierarchy, allometry, and entropy. Such concepts enable a spatial scientist to transcend the restrictions imposed upon the discovery of general principles both within geography and, more broadly, within the framework of mankind's organized system of knowledge. These restrictions, as noted previously, are a function of the emphasis upon the non-spatial properties of phenomena. Through this route of generality and abstraction, it has been argued, lies the way to a general science of spatial systems.

A number of common threads may be identified throughout the view of spatial process that has been developed in the preceding two chapters. The concepts of entropy, of a maximally probable state, of least effort, and of competition assume principal roles in both growth and organization. Further, these concepts reflect a tendency on the part of spatial systems toward structures that are optimal or maximally efficient with respect to a specific set of criteria. Hierarchical order and allometric growth (a hierarchy that increases so as to maintain a fixed bifurcation ratio or ratio of areas may be said to grow allometrically) are the common spatial manifestations of this optimality in a social–economic system and a physical environment where flows between components must be maintained and where the costs of these flows are neither negligible nor prohibitive. Figure 9.1 suggests possible optima in systems where: (a) each element is not dependent upon interaction with any other element and exploits uniformly distributed resources, and (b) the cost of interaction is extremely high but necessary for the survival of each element, and resources are concentrated.

It is not suggested that all systems are optimal but, rather, that self-organizing systems generally tend to proceed from less optimal toward more optimal forms. There appears to exist some natural selection phenomenon in spatial systems. The optimal and, thus, more efficient systems are the ones that are best able to maintain themselves and to undergo further development.

Geography, as all other branches of science which purport to study some aspect of existence in the real world, must be able to incorporate the notions of development and change somewhere within its framework. As noted in chapter 1, however, the discipline has always had difficulty in dealing with process. Schaefer's (264) view that geography is essentially morphological, containing no reference to time and change and dependent upon the 'process laws' developed in other disciplines, is well known. The concept of spatial process proposed here and examined in some detail in two of its identifiable and related aspects, growth and organization, eliminates some of the difficulty by holding time independent, in the manner with which most studies of process treat space. Thus, change in the spatial properties of objects and events becomes the primary focus of the concept. Not only is this a more tractable problem for geographers to investigate but it may also represent a means of eliminating the form–process distinction which Schaefer helped to establish.

In the preceding discussions of growth and organization it has been

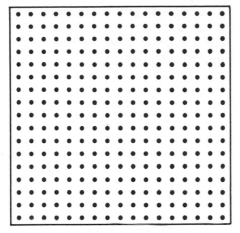

A. Independent elements exploiting uniformly distributed resources.

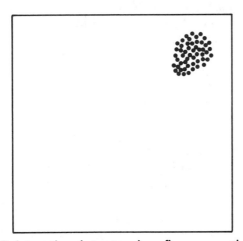

B. Interacting elements, where flows are costly.

Figure 9.1 *Optimal structure of systems under alternative conditions*

much easier to identify and to examine the various manifestations of spatial process – morphology, structure, shape, spatial hierarchy, connectivity, areal proportions, branching angles, bifurcation ratios, and so forth – than to establish the specific mechanisms of growth and organization which bring these about. That is to say, the discussion of spatial process has repeatedly reverted to the manifestations of change, leaving the manner of transition between forms largely untreated except for references to optimality, equilibrium, growth gradients and so forth. It is possible to argue, therefore, that we have not been discussing spatial process at all, and certainly not 'process' in its most rigorous sense. It is true that the usual social, economic, cultural, and physical phenomena which are conventionally subsumed under the term process have not been given detailed consideration. Before categorically rejecting the term spatial process and the utility of the concept, however, two parallel cases need to be explored.

First, the spatial process approach has distinct similarities with the conventional treatment of process in geography. The latter involves the attempt to identify a mathematical model which will describe the transition between one spatial form and another or, more precisely, to test the hypothesis that a certain mathematical model accounts for a certain pattern of points, lines or areas. Although it is recognized that changes are the result of forces (social, economic, or otherwise), this form of inquiry is 'not directly concerned with the forces, but with the changes' (116, p. 76). Tinkler's (311) discussion of matrix operations as processes in graphs and the treatment by Getis and Boots (118) of *Models of Spatial Processes* represent recent examples of a viewpoint which tends to regard a mathematical operation or a stochastic model as process. Like the spatial process formulation the mathematical model approach does not attempt to treat the behavioral or physical phenomena underlying change in spatial form. Unlike the latter, however, the former does attempt to determine explanatory factors or principles, even if these do seem overly abstract or ambiguous.

Second, there is a distinct parallel between the spatial process formulation that has been proposed for geography and the situation in the field of biology. Work by morphogenetic researchers has conclusively demonstrated that the difference between an arm and a leg, or between a man and a chimpanzee, is a difference in the spatial organization, the relative position, of a corresponding set of differential cell types. That is, the key to the differential organization of various forms of life does not lie in the cells as such; it lies in how these basic units are arranged in space during development (373, p. 154). Further, the distinct possibility exists that this type of positional information may have underlying universality; gradients in positional information may underlie pattern formation in an enormous variety of organisms.

Although biologists have determined rules governing the succession of forms in living beings during their development, the molecular processes which underlie these rules are completely unknown.

> It is sometimes held that no real progress has been made until a biological mechanism is placed on a firm molecular basis (until, in this case, the molecular nature of the gradients, or positional signals, is known). Such a view denies the existence of different levels of organization at which one can meaningfully investigate biological processes. Developmental biologists would like to know the molecular basis of pattern formation, but at present there is no obvious way to find it. . . . Perhaps we should be less apologetic and remember that. . . *unless we have the right phenomenology we do not know what we are trying to explain or where to look for the explanation* (373, p. 164; emphasis added).

The geographer, like the biologist, is often confronted by the question of why certain spatial arrangements exist and, similarly, is unable to specify the mechanisms by which the pattern is established, although he may have the capability to describe, predict or even, by reference to general concepts such as optimality, minimization or equilibrium, furnish an explanation-in-principle for morphology.

It may well be the case that, as *lex parsimonae* would indicate, broadly applicable explanation-in-principle in which general rules governing the succession of forms are established is more appropriate to the goals of theoretical science (see chapter 1) than a detailed understanding of the complex set of variables and interrelationships which constitute the mechanisms of change. This notion suggests that a parallel exists between the concept of spatial process and the substance referents of relational space (see chapter 2).[1] Just as the substance referents which describe relational space may (or, really, *must*) be ignored in order to generalize and to render operational the investigation of spatial problems (242), so perhaps must the mechanism which connects spatial forms. Although the rigorous usage of process, in the sense of the specification of both the sequence of events and the mechanism of change, is not satisfied by the approach outlined in the two preceding chapters, the interaction between spatial structure and movement in space and the corresponding set of principles represent sufficiently dynamical properties so that the term spatial process is justified. For linguistic purists, however, it is possible to borrow from developmental biology the term used to describe inquiry in which the fundamental question of the mechanism underlying spatial pattern cannot be established – morphodynamics (125).

Schaefer may have been correct; geography may be an essentially morphological discipline which, on its own, cannot adequately deal with process in its strict sense. The preceding chapters have not furnished strong evidence with which to dispute his view. Geography, however, is not alone in this predicament. As noted above, biology, perhaps the discipline in which development, transition and dynamism occupy the most central position, is similarly deficient; many other disciplines may likewise be judged to be inadequate from this criterion. This suggests that, given man's level of knowledge about himself and his environment, a thorough understanding of process, in its most rigorous sense, is not yet possible except in several of the physical sciences. The general spatial systems approach, with its eclectic and trans-disciplinary emphasis, may provide valuable insights to this ongoing search for understanding.

II. The Systems Approach in Geography

The systems approach, under its varied guises, has functioned as an organizing framework within the discipline of geography, providing a conceptual and empirical focus to those researchers who have found some merit in one or more aspects of the formulation. As noted in chapter 1, however, the approach is actually composed of three distinguishable elements: systems analysis, with its emphasis upon the control of specific systems; systems theory, stressing the identification and operation of system elements and properties; and general systems theory, serving primarily as a philosophical and epistemological basis for a style of inquiry. Although these three elements overlap to a considerable extent, each represents a fundamentally different method of approaching the search for knowledge.

The explicit manifestation of the systems framework in the discipline progressed from an amorphous notion which geographers were exhorted to utilize to a conceptual formulation which facilitated our understanding of the world, and thence to substantive operational methodologies for the analysis of spatial problems. These explicit attempts to implement a systems approach in geography were preceded, however, by a number of formulations which implicitly adopted the basic tenets of systems thinking. The demonstrated utility of the central place, potential, and dimensional approaches helped to create the intellectual climate in which the formal introduction of systems thinking was accomplished.

The discipline of geography does not exist or function in isolation; rather it contributes to and benefits from the general knowledge acquired through the human enterprise known as science. The systems approach was introduced into geography, implicitly at first and then explicitly, because of the prevailing intellectual climate in science at mid-century. More specifically, it was introduced into geography in an attempt to focus the methodological basis of the discipline upon those sets of elements and relationships which science was finding particularly useful in attempting to understand the real world.

As Kuhn (174) has pointed out, the progress of science is marked by the occurrence of revolutions or paradigm shifts. The end of a scientific revolution is signalled when a new paradigm becomes part of the accepted wisdom. Burton (58), for example, stated in 1963 that the 'quantitative revolution' in geography, which had commenced in the early 1950s, had ended; the once revolutionary methods of spatial analysis which had been introduced into the discipline had, by 1963, become part of the conventional wisdom of the field. Both in science generally and in the discipline of geography the systems 'revolution' appears to have ended. The once novel concepts of holism, interrelatedness, complexity, feedback, counter-intuitive behavior, and so forth are now generally acknowledged to represent valid aspects of the real world. Thus, it is somewhat difficult at the present time to identify analytical (as distinct from humanistic/phenomenological) research in geography which does not incorporate systems thinking, at least implicitly. Perhaps Chisholm (65) was correct when he observed that general systems theory is only a formal name for common sense.

Each of the three aspects of the systems approach that have been identified has some degree of utility for the analysis of spatial problems. Systems analysis, largely isomorphic to operations research, is of considerable practical significance in solving problems of efficient structure and/or movement in spatial systems. Since this approach has direct application to concrete problems, it has received much attention. Systems theory represents not only a useful set of concepts for specifying the interrelations between a set of elements but also a useful set of principles for modelling the behavior of those elements. One or more of the systems theoretic concepts has direct application to most aspects of spatial analysis. In addition, the number of text books and courses that have employed these concepts suggests that they possess considerable pedagogic utility. General systems theory, although farthest removed from direct operational utility, is a valuable methodological element of theoretical science. The search for meaningful commonalities between various types of spatial systems must be regarded as a worthwhile endeavor if one accepts the goals of theoretical science.

Each of these three elements of systems thinking is incorporated in the general spatial systems approach. The major emphasis, however, is upon general systems theory, which furnishes the epistemological basis of the approach and provides a context for the tools and concepts of systems analysis and systems theory. It is to a final consideration of the general spatial systems approach that we turn in bringing this discussion to a conclusion.

III. The General Spatial Systems Approach

Parts Two and Three of this book have been concerned with amplifying upon the notion of general spatial systems theory, originally articulated by Warntz and given preliminary development both in his work and, implicitly, in that of a number of other authors writing in the *Harvard Papers in Theoretical Geography* series. An attempt has been made to provide some of the detail missing in the early formulations and to furnish a suitable focus for the approach – the notion of spatial process.

It must be recognized that this discussion represents but one interpretation of a general spatial systems approach; other interpretations with varying emphases are conceivable. The underlying purpose and philosophy of this attempt, however, does not depart from that expressed by Warntz: it is a 'valid and important intellectual endeavor' to explore the question of whether or not a certain (small) number of systematic spatial properties are evident in phenomena that may be judged to differ widely in their non-spatial aspects.

The rather wide-ranging intellectual excursion undertaken here – broadly organized around the concepts of spatial structure, movement in space, spatial process, growth and organization – has attempted to explore this question in some detail. The concepts, principles, and tools that have been touched upon have failed to positively verify that meaningful spatial commonalities exist over a wide range of phenomena. Verification is not an appropriate criterion, however, as science progresses through falsification. On the basis of the somewhat limited and necessarily selective set of evidence examined, there appears to be no basis for rejecting the proposition that a set of broadly applicable spatial properties may be identified. Further, the notion that these properties have applicability both within the conventional boundaries of geography and across the traditional boundaries of disciplines may not be rejected.

A common criticism of general systems approaches is that, although they represent intellectually valid activity, their usefulness is minimal. The general spatial systems approach may not be subject to falsification, but what real value does it have? Any evaluation of an intellectual viewpoint is dependent upon the set of criteria against which it is judged. Here the distinct but complementary standards of intellectual and operational utility are often employed with some relative weight being assigned to each. If operational utility, the degree of applicability to the solution of a specific problem, is employed as the primary criterion, the broad conceptual structure of the general spatial systems approach is likely to be of far less value than specific techniques. On the other hand, in terms of its utility in promoting human knowledge, in its broadest sense, such a general approach may be far more desirable than the specialized operational tools.

The fundamental distinction which must be made is between the various levels at which knowledge may be sought. The general spatial systems approach seems to have considerable utility at the higher, broader levels of inquiry and substantially less at the more specific levels. The question then becomes one of the level at which knowledge ought to be pursued. As noted in chapter 1, this issue has been the subject of considerable debate in science, especially between the General Systems theorists and those adopting a narrower, more tractable, interpretation of the goals of science. The only reasonable answer to the question is one which recognizes that each approach is complementary to the other and necessary to the accomplishment of the objectives of science. Therefore, it follows that the general spatial systems approach is both a useful viewpoint and a necessary element of a comprehensive examination of reality. In addition, the argument may be made that since it is more general it subsumes the narrower approaches.

One advantage of adopting a higher level approach is that it functions, in essence, as a framework, providing not just an algorithm for solving a problem but, more generally, indicating the range of concepts and properties that should be investigated. The general spatial systems approach may be regarded as a framework, both within the discipline of geography and, more generally, within science. As any framework, it has certain identifiable characteristics (see chapter 1). Two of these have already been touched upon. First, a framework cannot be proved or validated; and, second, it involves a value judgment by indicating what types of phenomena should come under investigation. The view that the higher levels of knowledge are inherently more worthy of examination, explicit in the general systems theoretic approach, is not without some basis, but clearly entails a value judgment.

Frameworks, in addition, are evolutionary in nature, being refined or discarded as new knowledge suggests more useful ways of organizing reality. The attempt here has been to refine Warntz's original formulation of general spatial systems theory. Finally, there is some advantage to be gained from employing a broad framework that explicitly relates developments in one area of knowledge to those in other areas. As we have seen, each field contributes to the advancement of science but, at the same time, the course of science as a whole determines the progress of its parts. The treatment of the general spatial systems approach has, it is hoped, been sufficiently broad to expose successfully the breadth of applicability of a small number of spatial properties and concepts, both within geography and across disciplinary boundaries.

There remains one fundamental problem with the general spatial systems approach that is not easily resolved. With its emphasis both upon the unity of the discipline, of science, and of nature, and upon the universality of certain properties and principles, the approach does not adequately consider whether or not the isomorphisms or homologies that are observed are meaningful ones, particularly when they cross phenomenological boundaries. In the comparison between human and physical systems, a critical distinction arises in that the former are self-conscious and generally considered to be goal-seeking while the latter operate according to the laws of physics. This is a basic philosophical problem, the implications of which will need to be clarified. Although the foregoing discussion of general spatial systemic properties has not considered this problem explicitly, the indication of much macroscopic-level research is that the collective behavior of a group of goal-seeking individuals possessing free will tend to follow certain principles which are sometimes formulated in a manner isomorphic to physical laws. Much of social science research tends to ignore this problem; thus, at present, it should not be regarded as an obstacle to attempts to develop a general spatial systems approach. In addition, there is further evidence that social–economic systems tend to be conservative ones which absorb anomalous behavior.

Greer-Wootten (131; 132) has stated that, in the narrow sense of systems analysis, general systems theory has considerable relevance for geography. This is largely beyond dispute, particularly among those acquainted with the range of tools that systems analysis furnishes to geography. It may be argued more broadly that the entire scope of tools, methods, and concepts that have been outlined under the rubric of the general spatial systems approach have relevance for geography. Although this viewpoint will not generally furnish neat answers to narrow questions, it does address itself to fundamentally important questions of higher levels of knowledge. For this reason alone the approach is worthy of continued pursuit.

Notes

[1] This parallelism was suggested by Professor R. McDaniel, written communication, May 1978.

References

1. ABBOTT, EDWIN A. *Flatland,* sixth edition. New York: Dover Publications, 1952.
2. ABLER, R., J. ADAMS, AND P. GOULD. *Spatial Organization: The Geographer's View of the World.* Englewood Cliffs, NJ: Prentice-Hall, 1971.
3. ACKERMAN, EDWARD A. *Geography as a Fundamental Research Discipline,* Research Paper No. 53, Department of Geography, University of Chicago, 1958.
4. ACKERMAN, EDWARD A. 'Where is a Research Frontier?' *Annals,* Association of American Geographers, LIII, 1963, 429–40.
5. ADLER, IRVING. *A New Look at Geometry.* New York: New American Library, 1966.
6. AJO, REINO. 'An Approach to Demographical System Analysis,' *Economic Geography,* XXXVIII, 1962, 359–71.
7. AMEDEO, DOUGLAS AND REGINALD G. GOLLEDGE. *An Introduction to Scientific Reasoning in Geography.* New York: John Wiley, 1975.
8. AMSON, J.C. 'Equilibrium and Catastrophic Modes of Urban Growth,' in E.L. Cripps (ed.), *Space Time Concepts in Urban and Regional Models,* London Papers in Regional Science, 4. London: Pion, 1974, 108–28.
9. ANUCHIN, V.A. 'Mathematization and the Geographic Method,' *Soviet Geography: Review and Translation,* XI, 1970, 71–81.
10. ANUCHIN, V.A. *Theoretical Problems of Geography.* Moscow: Geografgiz, 1960.
11. ANUCHIN, V.A. 'Theory in Geography,' in R. Chorley (ed.), *Directions in Geography.* London: Methuen, 1973, 43–63.
12. APPEL, KENNETH AND WOLFGANG HAKEN. 'The Solution of the Four-Color Map Problem,' *Scientific American,* October 1977, 108–21.
13. ASHBY, W.R. 'Design for a Brain,' *Electronic Engineering,* XX, 1948.
14. ATTNEAVE, F. AND M.D. ARNOULT. 'The Quantitative Study of Shape and Pattern Perception,' *Psychological Bulletin,* LIII, 1956, 452–71.
15. AUERBACH, F. 'Das Gesetz der Bevolkerungskonzentration,' *Petermann's Geographische Mitteilungen,* LIX, 1913.
16. BALKUS, KOZMAS. 'An Outline for the Theory of Man-Made Space: Essays in Urbanology, Number One,' *Harvard Papers in Theoretical Geography,* Paper 38, Harvard University, 1970.
17. BANNISTER, G. 'Towards a Model of Impulse Transmissions for an Urban System,' *Environment and Planning,* VIII, 1976, 385–94.
18. BARROWS, H.H. 'Geography as Human Ecology,' *Annals,* Association of American Geographers, XIII, 1923, 1–14.

19. BENNETT, R.J. AND R.J. CHORLEY. *Environmental Systems: Philosophy and Control.* London: Methuen, 1978.
20. BERRY, BRIAN J.L. 'Approaches to Regional Analysis: a Synthesis,' *Annals,* Association of American Geographers, LIV, 1964, 2–11.
21. BERRY, BRIAN J.L. 'Cities as Systems Within Systems of Cities,' *Papers of the Regional Science Association,* XIII, 1964, 147–63.
22. BERRY, BRIAN J.L. 'City Size Distributions and Economic Development,' *Economic Development and Cultural Change,* IX, 1961, 573–88.
23. BERRY, BRIAN J.L. 'Essays on Commodity Flows and the Spatial Structure of the Indian Economy,' Research Paper No. 11, Department of Geography, University of Chicago, 1966.
24. BERRY, BRIAN J.L. *Geography of Market Centers and Retail Distribution.* Englewood Cliffs, NJ: Prentice-Hall, 1967.
25. BERRY, BRIAN J.L. 'A Paradigm for Modern Geography,' in R. Chorley (ed.), *Directions in Geography.* London: Methuen, 1973, 3–21.
26. BERRY, BRIAN J.L. AND WILLIAM GARRISON. 'The Functional Bases of the Central Place Hierarchy,' *Economic Geography,* XXXIV, 1958, 145–54.
27. BERRY, BRIAN J.L. AND ALLEN PRED. *Central Place Studies.* Philadelphia: Regional Science Research Institute, 1965.
28. BERTALANFFY, LUDWIG VON. *General System Theory.* New York: Braziller, 1972.
29. BERTALANFFY, LUDWIG VON. 'General System Theory–A Critical Review,' in Walter Buckley (ed.), *Modern Systems Research for the Behavioral Scientist.* Chicago: Aldine, 1968, 11–30.
30. BERTALANFFY, LUDWIG VON. 'General Systems Theory: A New Approach to the Unity of Science,' *Human Biology,* XXIII, 1951, 303–61.
31. BIBBY, JOHN. 'Infinite Rivers and the Steinhaus Paradox,' *Area,* IV, 1972, 214.
32. BJORKLUND, E.M. AND ALLEN K. PHILBRICK. 'The Spatial Configuration of Mental Process,' Paper presented to the Second Conference on the Geography of the Future, University of Montreal, February, 1972.
33. BLACK, G. *The Application of Systems Analysis to Government Operations.* New York: Praeger, 1968.
34. BLAIR, D.H. AND T.H. BLISS. *The Measurement of Shape in Geography,* Bulletin of Quantitative Data for Geographers, No. 11, Department of Geography, University of Nottingham, 1967.
35. BLALOCK, H.M. AND ANN B. BLALOCK. 'Toward a Clarification of System Analysis in the Social Sciences,' *Philosophy of Science,* XXVI, 1959, 84–92.
36. BLAUT, J.M. 'Object and Relationship,' *The Professional Geographer,* XIV, 1962, 1–7.
37. BLAUT, J.M. 'Space and Process,' *The Professional Geographer,* XIII, no. 4, 1961, 1–7.
38. BOHR, NIELS. *Atomic Physics and Human Knowledge.* New York: 1958.
39. BON, RANKO. 'Allometry in Micro-Environmental Morphology,' *Harvard Papers in Theoretical Geography,* Paper E, Harvard University, 1972.
40. BOULDING, KENNETH E. 'General System Theory–The Skeleton of Science,' in Walter Buckley (ed.), *Modern Systems Research for the Behavioral Scientist.* Chicago: Aldine, 1968.
41. BOULDING, KENNETH E. 'Toward a General Theory of Growth,' *General Systems,* I, 1956, 66–75.

42. BOYCE, R. AND W. CLARK. 'The Concept of Shape in Geography,' *Geographical Review*, LIV, 1964, 561–72.

43. BRIDGMAN, P.W. *The Logic of Modern Physics*. New York: Macmillan, 1972.

44. BRIDGMAN, P.W. *The Nature of Physical Theory*. Princeton: Princeton University Press, 1936.

45. BRILLOUIN, L. *Science and Information Theory*, second edition. New York: Academic Press, 1962.

46. BROWN, LAWRENCE A. 'Diffusion Dynamics,' *Lund Studies in Geography*, Series B, XXIX, 1968.

47. BROWN, LAWRENCE A. *Diffusion Processes and Location*. Philadelphia: Regional Science Research Institute, 1968.

48. BROWN, LAWRENCE A. 'On the Use of Markov Chains in Movement Research,' *Economic Geography*, XLVI, 1970, 393–403.

49. BROWN, LAWRENCE A. AND ERIC G. MOORE. 'The Intra-Urban Migration Process: A Perspective,' in Larry S. Bourne (ed.), *Internal Structure of the City*. New York: Oxford University Press, 1971, 200–9.

50. BUCK, R.C. 'On the Logic of General Behavior Systems Theory,' *Minnesota Studies in the Philosophy of Science*, I, 1956, 223–38.

51. BUNGE, WILLIAM. 'Detroit Humanly Viewed,' in R.F. Abler, D.G. Janelle, A.K. Philbrick, and J.W. Sommer (eds), *Human Geography in a Shrinking World*. N. Scituate, Mass.: Duxbury Press, 1975, 149–81.

52. BUNGE, WILLIAM. 'Ethics and Logic in Geography,' in R. Chorley (ed.), *Directions in Geography*. London: Methuen, 1973, 317–31.

53. BUNGE, WILLIAM. 'Fred K. Schaefer and the Science of Geography,' *Harvard Papers in Theoretical Geography*, Paper A, Harvard University, 1968.

54. BUNGE, WILLIAM. 'Locations Are Not Unique,' *Annals*, Association of American Geographers, LVI, 1966, 375–6.

55. BUNGE, WILLIAM. *Theoretical Geography*, second edition. Lund: Gleerup, 1966.

56. BUNGE, WILLIAM AND R. BORDESSA. *The Canadian Alternative*, Geographical Monographs, no. 2, Atkinson College, York University, 1975.

57. BUNGE, WILLIAM *et al.* 'A Report to the Parents of Detroit on School Decentralization,' in P.W. English and R.C. Mayfield (eds), *Man, Space, and Environment*. New York: Oxford University Press, 1972, 499–533.

58. BURTON, IAN. 'The Quantitative Revolution and Theoretical Geography,' *The Canadian Geographer*, VII, 1963, 151–62.

59. CAREY, HENRY C. *Principles of Social Science*. Philadelphia: Lippincott, 1858–9.

60. CARROTHERS, GERALD A. 'An Historical Review of the Gravity and Potential Concepts of Human Interaction,' *Journal of the American Institute of Planners*, XXII, 1956, 94–102.

61. CAYLEY, ARTHUR. 'On the Colouring of Maps,' *Proceedings, Royal Geographical Society*, I, 1879, 259–61.

62. CAYLEY, ARTHUR. 'On Contour and Slope Lines,' *The London, Edinburgh, and Dublin Philosophical Magazine and Journal of Science*, XVIII, 1859, 264–8.

63. CHEIN, I. 'Veracity vs. Truth in Scientific Enterprise,' address to American Psychological Association, Sept. 1967.

64. CHAPMAN, G.P. *Human and Environmental Systems: A Geographer's Appraisal.* London: Academic Press, 1977.

65. CHISHOLM, MICHAEL. 'General Systems Theory and Geography,' *Transactions,* Institute of British Geographers, XLII, 1967, 45–52.

66. CHISHOLM, MICHAEL. *Human Geography: Evolution or Revolution?* Baltimore: Penguin Books, 1975.

67. CHORAFAS, D.N. *Systems and Simulation.* New York: Academic Press, 1965.

68. CHORLEY, RICHARD J. 'Geomorphology and General Systems Theory,' US Geological Survey, Professional Paper 500-B, 1962, 1–10.

69. CHORLEY, R.J. AND B.A. KENNEDY. *Physical Geography: A Systems Approach.* London: Prentice-Hall International, 1971.

70. CHRISTALLER, WALTER. *Central Places in Southern Germany,* English edition, translated by C.W. Baskin. Englewood Cliffs, NJ: Prentice-Hall, 1968.

71. CHURCHMAN, C.W. *The Systems Approach.* New York: Dell Publishing, 1968.

72. CLARK, P.J. AND F.C. EVANS. 'Distance to Nearest Neighbor as a Measure of Spatial Relationships in Populations,' *Ecology,* XXXV, 1954, 445–53.

73. CLAYTON, C. 'Communication and Spatial Structure,' *Tijdschrift voor Economische en Sociale Geografie,* LXV, 1974, 271–7.

74. CLIFF, A. 'The Neighborhood Effect in the Diffusion of Innovation,' *Transactions,* Institute of British Geographers, XLII, 1968, 75–84.

75. COFFEY, WILLIAM. 'Allometric Growth in Urban and Regional Social–Economic Systems,' *Canadian Journal of Regional Science,* II, 1979, 49–65.

76. COFFEY, WILLIAM. 'Geographic Education in the Canadian University: Is This What We Really Want?' *Ontario Geography,* no. 11, 1977, 45–55.

77. COFFEY, WILLIAM. 'Income Relationships in Boston and Toronto: A Tale of Two Countries?,' *The Canadian Geographer,* XXII, 1978, 112–29.

78. COFFEY, WILLIAM. 'A Macroscopic Analysis of Income Regions in Metropolitan Boston,' *The Professional Geographer,* XXIX, 1977, 40–6.

79. COFFEY, WILLIAM. 'The Spatial Configuration of Social–Economic Integration: Metropolitan Boston,' unpublished MA Thesis, Department of Geography, University of Western Ontario, 1974.

80. COLE, J.P. AND C.A.M. KING. *Quantitative Geography.* London: John Wiley, 1968.

81. CROWE, P.R. 'On Progress in Geography,' *The Scottish Geographical Magazine,* LIV, no. 1, 1938, 1–19.

82. CURRY, LESLIE. 'Central Places in the Random Spatial Economy,' *Journal of Regional Science,* VII, 1967, 217–38.

83. DACEY, M.F. 'Analysis of Central Place and Point Patterns by a Nearest Neighbor Method,' *Lund Studies in Geography,* Series B, XIV, 1962, 55–75.

84. DACEY, M.F. 'Description of Line Patterns,' *Northwestern University Studies in Geography,* XIII, 1967, 277–87.

85. DACEY, M.F. 'Modified Poisson Probability Law for Point Patterns More Regular Than Random,' *Annals,* Association of American Geographers, LIV, 1964, 559–65.

86. DACEY, M.F. 'A Review of Measures of Contiguity for Two and K-Color Maps,' in B.J.L. Berry and D.F. Marble (eds), *Spatial Analysis.* Englewood Cliffs, NJ: Prentice-Hall, 1968.

87. DACEY, MICHAEL F. 'Some Questions About Spatial Distributions,' in R.J. Chorley (ed.), *Directions in Geography*. London: Methuen, 1974, 127–51.
88. DAVIS, WILLIAM MORRIS. 'The Geographical Cycle,' *Geographical Journal*, XIV, 1899, 481–504.
89. DE GEER, STEN. 'On the Definition, Method and Classification of Geography,' *Geografiska Annaler*, V, 1923, 1–37.
90. DE SMITH, M.J. 'A Method for Analyzing Complex Bounded Shapes,' unpublished MA Thesis, Department of Geography, University of Western Ontario, 1974.
91. DEUTSCH, A.J. 'A Subway Named Mobius,' in Isaac Asimov (ed.), *Where Do We Go From Here?* Greenwich, Conn.: Fawcett, 1971.
92. DOLPHIN, VERNON. 'The Earth is the Home of Man,' unpublished PhD Thesis, Department of Education, Harvard University, 1970.
93. DUNN, E.S. JR. *The Location of Agricultural Production*. Gainesville, Fla.: University of Florida Press, 1954.
94. DUTTON, GEOFFREY H. 'Macroscopic Aspects of Metropolitan Evolution,' *Harvard Papers in Theoretical Geography*, Paper 1, Harvard University, 1970.
95. DUTTON, GEOFFREY. 'National and Regional Parameters of Growth and Distribution of Population in the United States, 1790–1970,' *Harvard Papers in Theoretical Geography*, Paper 5, Harvard University, 1971.
96. EASTON, D. *A Framework for Political Analysis*. Englewood Cliffs, NJ: Prentice-Hall, 1965.
97. EASTON, D. *A Systems Analysis of Political Life*. New York: John Wiley, 1965.
98. ECKHARDT, ROBERT C. 'Minimum Time Paths of the Arctic Tern,' *Harvard Papers in Theoretical Geography*, Paper 26, Harvard University, 1969.
99. EDDINGTON, A.S. *The Nature of the Physical World*. New York: Macmillan, 1943.
100. EICHENBAUM, JACK AND STEPHEN GALE. 'Form, Function, and Process: A Methodological Inquiry,' *Economic Geography*, XLVII, 1971, 525–44.
101. ELIOT HURST, MICHAEL E. 'The Geographic Study of Transportation, Its Definition, Growth, and Scope,' in M.E. Eliot Hurst (ed.), *Transportation Geography: Comments and Readings*. New York: McGraw-Hill, 1974, 1–15.
102. ELIOT HURST, M.E. *A Geography of Economic Behavior*. N. Scituate, Mass.: Duxbury Press, 1972.
103. ELSDALE, TOM. 'Pattern Formation in Fibroblast Cultures, an Inherently Precise Morphogenetic Process,' in C.H. Waddington (ed.), *Toward a Theoretical Biology*, vol. 4. Chicago: Aldine, 1972, 95–108.
104. ENTRIKIN, J. NICHOLAS. 'Contemporary Humanism in Geography,' *Annals*, Association of American Geographers, LXVI, 1976, 615–32.
105. EULER, LEONHARD. 'Solution of a Problem Belonging to the Geometry of Position,' St Petersburg Academy of Science, 1736.
106. EVES, HOWARD. *A Survey of Geometry*, vol. II. Boston: Allyn and Bacon, 1965.
107. FEIN, ELIHU. 'Demography and Thermodynamics,' *American Journal of Physics*, XXXVIII, no. 12, 1970, 1373–9.
108. FENNEMAN, NEVIN M. 'The Circumference of Geography,' *Annals*, Association of American Geographers, IX, 1919, 3–11.
109. FOOTE, DON C. AND BRYN GREER-WOOTTEN. 'An Approach to Systems

Analysis in Cultural Geography,' *The Professional Geographer,* xx, 1968, 86–91.

110. GARDNER, MARTIN. *Logic, Machines, and Diagrams.* New York: McGraw-Hill, 1958.

111. GARDNER, MARTIN. 'Mathematical Games,' *Scientific American,* CCXXXIV, no. 4, April 1976, 126–30.

112. GARRISON, WILLIAM. 'Connectivity of the Interstate Highway System,' *Papers and Proceedings, Regional Science Association,* VI, 1960, 121–37.

113. GEORGESCU-ROEGEN, NICHOLAS. *The Entropy Law and the Economic Process.* Cambridge, Mass.: Harvard University Press, 1971.

114. GERARD, RALPH W. 'Hierarchy, Entitation, and Levels,' in L.L. Whyte, A.G. Wilson, and D. Wilson (eds), *Hierarchical Structures.* New York: American Elsevier, 1969, 215–30.

115. GETIS, ARTHUR. 'The Determination of the Location of Retail Activities with the Use of a Map Transformation,' *Economic Geography,* XXXIX, 1963, 14–22.

116. GETIS, ARTHUR. 'Representation of Spatial Point Pattern Processes by Polya Models,' in M. Yeates (ed.), *Proceedings of the 1972 Meeting of the IGU Commission on Quantitative Geography.* Montreal: McGill–Queen's University Press, 1974, 76–100.

117. GETIS, ARTHUR. 'Temporal Analysis of Land Use Patterns with the Use of Nearest Neighbor and Quadrat Methods,' *Annals,* Association of American Geographers, LIV, 1964, 391–9.

118. GETIS, ARTHUR AND BARRY BOOTS. *Models of Spatial Processes.* Cambridge Geographical Studies, No. 8. Cambridge: Cambridge University Press, 1978.

119. GODDARD, J.B. 'Functional Regions within the City Centre: A Study of Factor Analysis of Taxi Flows in Central London,' *Transactions,* Institute of British Geographers, IL, 1970, 161–82.

120. GOLLEDGE, R.G. AND G. RUSHTON. *Multidimensional Scaling: Review and Geographical Applications,* Association of American Geographers Technical Paper No. 10, 1972.

121. GOOD, C.M. 'Periodic Markets and Travelling Traders in Uganda,' *Geographical Review,* LXV, 1975, 45–72.

122. GOODCHILD, MICHAEL F. 'Spatial Choice in Location–Allocation Problems: The Role of Endogenous Attraction,' *Geographical Analysis,* X, 1978, 65–72.

123. GOODCHILD, MICHAEL F., NINA SIU-NGAN LAM, AND JOHN D. RADKE. 'An Approach to the Study of Nodal Growth,' *Canadian Journal of Regional Science,* II, 1979, 67–76.

124. GOODCHILD, MICHAEL F. AND BRYAN H. MASSAM. 'Some Least-Cost Models of Spatial Administrative Systems in Southern Ontario,' *Geografiska Annaler,* LII-B, 1969, 86–94.

125. GORDON, RICHARD AND ANTONE G. JACOBSON. 'The Shaping of Tissues in Embryos,' *Scientific American,* CCXXXVIII, June 1978, 106–13.

126. GOULD, PETER. 'On Mental Maps,' Michigan Inter-University Community of Mathematical Geographers, Discussion Paper No. 9, 1966.

127. GOULD, PETER. 'Pedagogic Review,' *Annals,* Association of American Geographers, LXII, 1972, 689–700.

128. GOULD, PETER. *Spatial Diffusion,* Association of American Geographers, Commission on College Geography, Resource Paper No. 4, 1969.

129. GOULD, STEPHEN Jay. 'Allometry and Size in Ontogeny and Phylogeny,' *Biological Reviews,* XLI, 1966, 587–640.

130. GRAVELIUS, H. *Flusskunde band 1.* Berlin and Leipzig: F.G. Goschenesche Verlagshamolung, 1914.

131. GREER-WOOTTEN, BRYN. 'The Role of General Systems Theory in Geographic Research,' Discussion Paper No. 3, Department of Geography, York University, Toronto, 1972.

132. GREER-WOOTTEN, BRYN. 'Some Reflections on Systems Analysis in Geographic Research,' in H.M. French and J.-B. Racine (eds), *Quantitative and Qualitative Geography: la nécessité d'un Dialogue.* Ottawa: University of Ottawa Press, 1971, 151–74.

133. GROBSTEIN, CLIFFORD. 'Hierarchical Order and Neogenesis,' in H. Pattee (ed.), *Hierarchy Theory.* New York: Braziller, 1973, 29–47.

134. GUELKE, LEONARD. 'Problems of Scientific Explanation in Geography,' *The Canadian Geographer,* XV, no. 1, 1971, 38–53.

135. GUYOT, RICHARD. 'Two Theorems for Geography,' in J. Nystuen (ed.), 'The Philosophy of Maps,' Michigan Inter-University Community of Mathematical Geographers, Discussion Paper No. 12, 1968, 66–76.

136. HACK, J.T. 'Studies of Longitudinal Stream Profiles in Virginia and Maryland,' US Geological Survey, Professional Paper 294-B, 1957.

137. HAGEN, E.E. *On the Theory of Social Change.* Homewood, Ill.: Dorsey Press, 1962.

138. HAGERSTRAND, TORSTEN. 'Aspects of the Spatial Structure of Social Communication and the Diffusion of Information,' *Papers of the Regional Science Association,* XVI, 1966, 27–42.

139. HAGERSTRAND, TORSTEN. *Diffusion of Innovations,* translated by Allan Pred. Chicago: The University of Chicago Press, 1968.

140. HAGGETT, PETER. *Locational Analysis in Human Geography.* New York: St Martin's Press, 1966.

141. HAGGETT, PETER AND RICHARD CHORLEY. *Network Analysis in Geography.* London: Edward Arnold, 1969.

142. HALL, EDWARD. *The Hidden Dimension.* Garden City, NY: Doubleday and Co., 1969.

143. HALL, A.D. AND R.E. FAGEN. 'Definition of System,' in Walter Buckley (ed.), *Modern Systems Research for the Behavioral Scientist.* Chicago: Aldine, 1968, 81–92.

144. HARTSHORNE, RICHARD. 'Exceptionalism in Geography Re-examined,' *Annals,* Association of American Geographers, XLV, 1955, 205–44.

145. HARTSHORNE, RICHARD. *The Nature of Geography.* Lancaster, Penn.: Association of American Geographers, 1939.

146. HARTSHORNE, RICHARD. *Perspective on the Nature of Geography.* Chicago: Rand McNally and Co., 1959.

147. HARVEY, DAVID. *Explanation in Geography.* New York: St Martin's Press, 1969.

148. HARVEY, DAVID. 'Geographical Processes and Point Patterns: Testing Models of Diffusion by Quadrat Sampling,' *Transactions,* Institute of British Geographers, XL, 1966, 81–95.

149. HARVEY, DAVID. 'Models of Spatial Patterns in Human Geography,' in R.J. Chorley and P. Haggett (eds), *Models in Geography.* London: Methuen, 1967, 549–608.

150. HARVEY, M.E., R.T. HOCKING AND J.R. BROWN. 'The Chromatic Travelling-Salesman Problem and its Application to Planning and Structuring Geographic Space,' *Geographical Analysis*, VI, 1974, 33–52.

151. HAYNES, ROBIN M. 'Dimensional Analysis: Some Applications in Human Geography,' *Geographical Analysis*, VII, no. 1, 1975, 51–67.

152. HEMPEL, CARL G. *Philosophy of Natural Science*. Englewood Cliffs, NJ: Prentice-Hall, 1966.

153. HEWITT, KENNETH AND F. KENNETH HARE. *Man and Environment*. Commission on College Geography, Resource Paper No. 20, Association of American Geographers, Washington, DC, 1973.

154. HORTON, R.E. 'Erosional Development of Streams and Their Drainage Basins: Hydrophysical Approach to Quantitative Morphology,' *Geological Society of America Bulletin*, LVI, 1945, 275–370.

155. HOTELLING, HAROLD. 'Stability in Competition,' *The Economic Journal*, XXXIX, 1929, 41–57.

156. HOWIE, GORDON. 'A Study of Rivers and Other Branching Systems,' *Harvard Papers in Theoretical Geography*, Paper 20, Harvard University, 1968.

157. HUDSON, JOHN C. 'Diffusion in a Central Place System,' *Geographical Analysis*, I, 1969, 45–58.

158. HUNTLEY, H.E. *Dimensional Analysis*. New York: Dover Publications, 1967.

159. HUREWICZ, WITHOLD AND HENRY WALLMAN. *Dimension Theory*. Princeton, NJ: Princeton University Press, 1941.

160. HUXLEY, JULIAN. *Problems of Relative Growth*, second edition. New York: Dover Publications, 1972.

161. HUXLEY, JULIAN. 'Relative Growth and Form Transformation,' *Proceedings of the Royal Society of London*, B137, 1950, 465–9.

162. IPSEN, D.C. *Units, Dimensions, and Dimensionless Numbers*. New York: McGraw-Hill, 1960.

163. ISARD, WALTER. *Methods of Regional Analysis: An Introduction to Regional Science*. New York: John Wiley, 1960.

164. KANSKY, K.J. 'Structure of Transport Networks: Relationships Between Network Geometry and Regional Characteristics,' Research Paper No. 84, Department of Geography, University of Chicago, 1963.

165. KAPLAN, M.A. 'International Systems,' in *International Encyclopedia of the Social Sciences*. New York: Macmillan and Free Press, 1968, 479–86.

166. KARLIN, ANDREW. 'Shapes as a Group,' in J. Nystuen (ed.), 'The Philosophy of Maps,' Michigan Inter-University Community of Mathematical Geographers, Discussion Paper No. 12, 1968.

167. KATES, R.W. 'Links Between Physical and Human Geography: A Systems Approach,' in *Introductory Geography: Viewpoints and Themes*, Commission on College Geography, Publication No. 5, Association of American Geographers, Washington, DC, 1967, 23–30.

168. KEMENY, JOHN G. *A Philosopher Looks at Science*. Princeton, NJ: D. Van Nostrand Co., 1959.

169. KHAILOV, K.M. 'The Orderliness of Biological Systems,' *General Systems*, XII, 1967, 29–37.

170. KING, LESLIE J. 'A Quantitative Expression of the Pattern of Urban Settlements in Selected Areas of the United States,' *Tijdschrift voor Economische en Sociale Geografie*, LIII, 1962, 1–7.

References

171. KING, L., E. CASETTI, AND D. JEFFREY. 'Economic Impulses in a Regional System of Cities: A Study of Spatial Interaction,' *Regional Studies,* III, 1969, 213–18.

172. KLIR, J. AND M. VALACH. *Cybernetic Modelling,* English edition. London: Iliffe Books, 1967.

173. KOESTLER, ARTHUR. *The Ghost in the Machine.* London: Pan, 1967.

174. KUHN, THOMAS S. *The Structure of Scientific Revolutions.* Chicago: University of Chicago Press, 1970.

175. LAMARCHE, R.H. 'A Flowgraph Technique for Measuring Urban System Characteristics,' paper presented at the Annual Meeting of the Association of American Geographers, New Orleans, April 1978.

176. LANGBEIN, W.B. AND L.B. LEOPOLD. 'River Meanders – Theory of Minimum Variance,' US Geological Survey, Professional Paper 422-H, 1966.

177. LANGTON, JOHN. 'Potentialities and Problems of Adopting a Systems Approach to the Study of Change in Human Geography,' in C. Board *et al.* (eds), *Progress in Geography,* vol. 4. London: Arnold, 1972, 125–79.

178. LASZLO, ERVIN. *Introduction to Systems Philosophy.* New York: Harper and Row, 1973.

179. LASZLO, ERVIN (ed.). *The Relevance of General Systems Theory.* New York: Braziller, 1972.

180. LASZLO, ERVIN. *The Systems View of the World.* New York: Braziller, 1972.

181. LAYZER, DAVID. 'The Arrow of Time,' *Scientific American,* CCXXXIII, no. 6, December 1975, 56–9.

182. LEE, D. AND G. SALLEE. 'A Method of Measuring Shape,' *Geographial Review,* LX, 1970, 555–63.

183. LEOPOLD, LUNA B. 'Rivers,' *American Scientist,* L, 1962, 511–37.

184. LEOPOLD, LUNA B. AND WALTER B. LANGBEIN. 'The Concept of Entropy in Landscape Evolution,' *Theoretical Papers in the Hydrologic and Geomorphic Sciences.* US Geological Survey, Professional Paper 500-A, 1962, 1–20.

185. LEOPOLD, L.B. AND T. MADDOCK. 'The Hydraulic Geometry of Stream Channels and Some Physiographic Implications,' US Geological Survey, Professional Paper 252, 1953.

186. LEOPOLD, L.B., M.G. WOLMAN, AND J.P. MILLER. *Fluvial Processes in Geomorphology.* San Francisco: Freeman, 1964.

187. LEVINS, RICHARD. *Evolution in Changing Environments: Some Theoretical Explorations.* Princeton, NJ: Princeton University Press, 1968.

188. LEVINS, RICHARD. 'The Limits of Complexity,' in H. Pattee (ed.), *Hierarchy Theory.* New York: Braziller, 1973, 109–27.

189. LEWIN, KURT. *Principles of Topological Psychology.* New York: McGraw-Hill, 1946.

190. LOSCH, AUGUST. *The Economics of Location,* second edition. New York: John Wiley, 1967.

191. LOWE, JOHN C. AND S. MORYADAS. *The Geography of Movement.* Boston: Houghton Mifflin, 1975.

192. LOWENTHAL, DAVID. 'Geography, Experience, and Imagination: Towards a Geographical Epistemology,' *Annals,* Association of American Geographers, LI, no. 3, 1961, 241–60.

193. LUBOWE, JOAN K. 'Stream Junction Angles in the Dendritic Drainage Pattern,' *American Journal of Science,* CCLXII, 1964, 325–39.

194. LUKERMAN, FRED. 'Geography: de facto or de jure,' *Journal of the Minnesota Academy of Science,* XXXII, 1965, 189–96.

195. LUKERMAN, FRED. 'Toward a More Geographic Economic Geography,' *Professional Geographer,* X, 1958, 2–10.

196. LYNCH, KEVIN. *The Image of the City.* Cambridge, Mass.: MIT Press, 1960.

197. MABOGUNJE, AKIN L. 'Systems Approach to a Theory of Rural–Urban Migration,' *Geographical Analysis,* II, 1970, 1–18.

198. McDANIEL, ROBERT AND M.E. ELIOT HURST. *A Systems Analytic Approach to Economic Geography.* Commission on College Geography, Publication No. 8, Association of American Geographers, Washington, DC, 1968.

199. MACKAY, J. ROSS. 'The Interactance Hypothesis and Boundaries in Canada: A Preliminary Study,' *The Canadian Geographer,* no. 11, 1958, 1–8.

200. MACKINNON, ROSS D. 'Dynamic Programming and Geographical Systems,' *Economic Geography,* XLVI, no. 2 (supplement), 1970, 350–66.

201. MACKINNON, ROSS D. AND GERALD M. BARBER. 'Optimization Models of Transportation Network Improvement,' *Progress in Human Geography,* I, 1977, 387–412.

202. MANDELBROT, BENOIT B. *Fractals: Form, Chance, and Dimension.* San Francisco: W.H. Freeman, 1977.

203. MARCH, LIONEL AND PHILIP STEADMAN. *The Geometry of Environment.* Cambridge, Mass.: MIT Press, 1974.

204. MARCHAND, BERNARD. 'Information Theory and Geography,' *Geographical Analysis,* IV, no. 3, 1972, 234–57.

205. MARTIN, R.L., N.J. THRIFT, AND R.J. BENNETT (eds). *Towards the Dynamic Analysis of Spatial Systems.* London: Academic Press, 1979.

206. MARUYAMA, MAGOROH. 'The Second Cybernetics: Deviation Amplifying Mutual Causal Processes,' *Cybernetica,* VI, 1963, 5–23.

207. MASSAM, BRYAN H. *Location and Space in Social Administration.* London: Arnold, 1975.

208. MAXWELL, JAMES CLERK. 'On Hills and Dales,' *The London, Edinburgh, and Dublin Philosophical Magazine and Journal of Science,* XL, 1870, 421–7.

209. MEDVEDKOV, Y. 'Entropy: An Assessment of Potentialities in Geography,' *Economic Geography,* XLVI, no. 2 (supplement), 1970, 306–16.

210. MESAROVIC, M.D. (ed.). *Views on General System Theory.* New York: John Wiley, 1964.

211. MORISAWA, M.E. 'Relation of Quantitative Geomorphology to Stream Flow in Representative Water-sheds of the Appalachian Plateau Province,' Office of Naval Research–Columbia University Department of Geology, Technical Report No. 20, 1959.

212. MORRILL, RICHARD L. 'The Negro Ghetto: Problems and Alternatives,' *Geographical Review,* LV, no. 3, 1965, 339–61.

213. MUELLER, J.E. 'An Introduction to the Hydraulic and Topographic Sinuosity Indexes,' *Annals,* Association of American Geographers, LVII, 1968, 371–85.

214. MURRAY, CECIL. 'The Physiological Principle of Minimum Work Applied to the Angle of Branching of Arteries,' *Journal of General Physiology,* IX, 1926, 835–41.

215. MUSHAM, H.V. 'Internal Migration in Open Populations,' in J. Sutter (ed.), *Human Displacements.* Monaco: Entretiens de Monaco, 1963.

216. NAGEL, E. *The Structure of Science*. New York: Harcourt, Brace & World, 1961.

217. NAROLL, RAOUL AND LUDWIG VON BERTALANFFY. 'The Principle of Allometry in Biology and the Social Sciences,' *Ekistics*, xxxvi, 1973, 244–52.

218. NATIONAL ACADEMY OF SCIENCES–NATIONAL RESEARCH COUNCIL. *The Science of Geography*. Washington, DC: NAS–NRC, 1965.

219. NEWLING, BRUCE E. 'Urban Growth and Spatial Structure: Mathematical Models and Empirical Evidence,' *The Geographical Review*, lvi, 1966, 213–25.

220. NORDBECK, STIG. 'The Law of Allometric Growth,' Michigan Inter-University Community of Mathematical Geographers, Discussion Paper No. 7, 1965.

221. NORDBECK, STIG. 'Urban Allometric Growth,' *Geografiska Annaler*, 53B, 1971, 54–67.

222. NYE, J.F. 'The Distribution of Stress and Velocity in Glaciers and Ice Sheets,' *Proceedings of the Royal Society of London*, A239, 1970, 113–33.

223. NYSTUEN, JOHN. 'Effects of Boundary Shape and the Concept of Local Convexity,' in John Nystuen (ed.), Michigan Inter-University Community of Mathematical Geographers, Discussion Paper No. 10, 1966.

224. NYSTUEN, JOHN D. AND MICHAEL F. DACEY. 'A Graph Theory Interpretation of Nodal Regions,' *Papers and Proceedings*, Regional Science Association, vii, 1961, 29–42.

225. OLSSON, GUNNAR. 'Central Place Systems, Spatial Interaction and Stochastic Processes,' *Papers*, Regional Science Association, xviii, 1966, 13–45.

226. OLSSON, GUNNAR. 'The Dialectics of Spatial Analysis,' *Antipode*, 1974, 50–62.

227. OLSSON, GUNNAR. *Distance and Human Interaction*. Philadelphia: Regional Science Research Institute, 1965.

228. *Oxford English Dictionary*. Compact Edition. New York: Oxford University Press, 1971.

229. PARETO, V. *Cours de l'Economie Politique*. (2 vols.) Lausanne: Rouge, 1896–7.

230. PATTEE, HOWARD. 'Physical Conditions for Primitive Functional Hierarchies,' in L.L. Whyte, A.G. Wilson and D. Wilson (eds), *Hierarchical Structures*. New York: American Elsevier, 1969, 161–77.

231. PATTEE, HOWARD. 'Unsolved Problems and Potential Applications of Hierarchy Theory,' in H. Pattee (ed.), *Hierarchy Theory*. New York: Braziller, 1973, 129–56.

232. PERKAL, JULIAN. 'On the Epsilon Length,' *Bulletin of the Polish Academy of Science*, C1, III, iv, no. 7, 1956, 399–403.

233. PERKAL, JULIAN. 'On the Length of Empirical Curves,' in John Nystuen (ed.), Michigan Inter-University Community of Mathematical Geographers, Discussion Paper No. 10, 1966.

234. PHILIPS, O.M. *The Heart of the Earth*. San Francisco: Cooper and Co., 1968.

235. PIAGET, J. 'How Children Form Mathematical Concepts,' *Scientific American*, clxxxix, November 1953.

236. PIAGET, J. AND B. INHELDER. *The Child's Conception of Space*. London: Routledge and Kegan Paul, 1956.

237. PLATT, J. 'Hierarchical Growth,' *Bulletin of the Atomic Scientists*, xv, 1970, 46–8.

238. PLATT, JOHN. 'Hierarchical Restructuring,' *General Systems*, xv, 1970, 49–54.

239. POINCARÉ, H. *Science and Hypothesis*. New York: Dover, 1952.

240. POINCARÉ, H. *The Value of Science*. New York: Science Press, 1929.

241. POOLER, JAMES A. 'The Origins of the Spatial Tradition in Geography: An Interpretation,' *Ontario Geography*, no. 11, 1977, 56–83.

242. POOLER, JAMES A. 'The Relational Concept of Space and Empirical Usefulness – A Comment,' *Geographical Analysis*, viii, 1976, 474–9.

243. PRIGOGINE, I. *Étude thermodynamique des phénomènes irreversibles*. Paris: Dunod, 1947.

244. RAPOPORT, ANATOL. 'Foreword,' in Walter Buckley (ed.), *Modern Systems Research for the Behavioral Scientist*. Chicago: Aldine, 1968.

245. RAPOPORT, ANATOL. 'The Search for Simplicity,' in Ervin Laszlo (ed.), *The Relevance of General Systems Theory*. New York: Braziller, 1972.

246. RASHEVSKY, N. 'Topology and Life: In Search of General Mathematical Principles in Biology and Sociology,' *General Systems*, i, 1956, 123–38.

247. RAVENSTEIN, E.G. 'The Laws of Migration,' *Journal of the Royal Statistical Society*, 48, 1885, 167–235.

248. RAY, D. MICHAEL. 'The Allometry of Urban and Regional Growth,' Paper presented to the IGU Commission on Regional Aspects of Economic Development, London, Canada, 1972.

249. RAY, D. MICHAEL, P.Y. VILLENEUVE, AND R.A. ROBERGE. 'Functional Prerequisites, Spatial Diffusion, and Allometric Growth,' *Economic Geography*, L, 1974, 341–51.

250. REILLY, W.J. *The Law of Retail Gravitation*. New York: W.J. Reilly Co., 1931.

251. REILLY, W.J. *Methods for the Study of Retail Relationships*. University of Texas, Bureau of Business Research, Research Monograph No. 4 (Bulletin 2994), 1929.

252. REYNOLDS, OSBORNE. 'An Experimental Investigation of the Circumstances which Determine Whether the Motion of Water Shall Be Direct or Sinuous, and of the Law of Resistance in Parallel Channels,' *Philosophical Transactions, Royal Society*, CLCCIV, 1883.

253. RICHARDSON, LEWIS F. 'The Problem of Contiguity,' *General Systems*, vi, 1961, 139–87.

254. RICHARDSON, MOSES. *Fundamentals of Mathematics*. New York: Macmillan, 1958.

255. ROBERTS, F.S. AND P. SUPPES. 'Some Problems in the Geometry of Visual Perception,' *Synthèse*, xvii, 1967, 173–201.

256. ROSEN, ROBERT. *Optimality Principles in Biology*. New York: Plenum Press, 1967.

257. RUECHARDT, EDUARD. *Light*. Ann Arbor: University of Michigan Press, 1958.

258. RUSSWURM, LORNE H. AND EDWARD SOMMERVILLE. *Man's Natural Environment*. N. Scituate, Mass.: Duxbury Press, 1974.

259. SACK, ROBERT DAVID. 'Chorology and Spatial Analysis,' *Annals*, Association of American Geographers, LXIV, no. 3, 1974, 439–52.

260. SACK, ROBERT DAVID. 'A Concept of Physical Space in Geography,' *Geographical Analysis*, V, no. 1, 1973, 16–34.
261. SACK, ROBERT DAVID. 'Geography, Geometry, and Explanation,' *Annals*, Association of American Geographers, LXII, no. 1, 1972, 61–78.
262. SACK, ROBERT DAVID. 'The Spatial Separatist Theme in Geography,' *Economic Geography*, L, no. 1, 1974, 1–19.
263. SAUER, CARL O. *Land and Life*. J.B. Leighley (ed.). Berkeley: University of California Press, 1963.
264. SCHAEFER, FRED K. 'Exceptionalism in Geography: A Methodological Examination,' *Annals*, Association of American Geographers, XLIII, no. 3, 1953, 226–49.
265. SCHEIDEGGER, A.E. 'On the Topology of River Nets,' *Water Resources Research*, III, 1967, 103–6.
266. SCHEIDEGGER, A.E. 'The Algebra of Stream-Order Numbers,' US Geological Survey Professional Paper 525-B, B187–9.
267. SCHUMM, S.A. 'Sinuosity of Alluvial Rivers on the Great Plains,' *Bulletin of the Geological Society of America*, LXXIV, 1963, 1089–1100.
268. SCOTT, ALLAN J. *Combinatorial Programming, Spatial Analysis and Planning*. London: Methuen, 1971.
269. SENIOR, M.L. AND A.G. WILSON. 'Explorations and Syntheses of Linear Programming and Spatial Interaction Models of Residential Location,' *Geographical Analysis*, VI, 1974, 209–38.
270. SHANNON, CLAUDE AND WARREN WEAVER. *The Mathematical Theory of Communication*. Urbana: University of Illinois Press, 1949.
271. SHOWERS, VICTOR. *The World in Figures*. New York: John Wiley, 1973.
272. SHREVE, R.L. 'Statistical Law of Stream Numbers,' *Journal of Geology*, LXXIV, 1966, 17–37.
273. SIMON, HERBERT A. 'On a Class of Skew Distribution Functions,' *Biometrica*, XLII, 1955, 425–40.
274. SIMON, HERBERT A. 'The Organization of Complex Systems,' in H. Pattee (ed.), *Hierarchy Theory*. New York: Braziller, 1973, 1–27.
275. SIMMONS, JAMES W. 'Changing Residence in the City: A Review of Intra-Urban Mobility,' *The Geographical Review*, LVIII, 1968, 622–51.
276. SMALLEY, IAN AND CLAUDIO VITA-FINZI. 'The Concept of "System" in the Earth Sciences, Particularly Geomorphology,' *Geological Society of America Bulletin*, LXXX, 1969, 1591–4.
277. SMART, J.S. AND A.J. SURKAN. 'The Relation Between Mainstream Length and Area in Drainage Basins,' *Water Resources Research*, III, 1967, 963–74.
278. SMITH, CYRIL STANLEY. 'Structural Hierarchy in Inorganic Systems,' in L.L. Whyte, A.G. Wilson and D. Wilson (eds), *Hierarchical Structures*. New York: American Elsevier, 1969, 61–85.
279. SMITH, R.H.T. 'Concepts and Methods in Commodity Flow Analysis,' *Economic Geography*, XLVI, 1970, 404–16.
280. SMITH, TERENCE R. 'Set-Determined Process and the Growth of Spatial Structure,' *Geographical Analysis*, VIII, 1976, 354–75.
281. SNOW, C.P. 'The Moral Unneutrality of Science,' *Science*, 133, 1961, 256 ff.
282. SOMMER, ROBERT. *Personal Space*. Englewood Cliffs, NJ: Prentice-Hall, 1969.

References

283. SPEIGHT, J.G. 'Meander Spectra of the Angabunga River,' *Journal of Hydrology*, III, 1965, 1–15.

284. STEINHAUS, HUGO. 'Length, Shape, and Area,' *Colloquium Mathematicum*, 1954, 1–13.

285. STEINHAUS, HUGO. *Mathematical Snapshots*. London: Oxford University Press, 1960.

286. STEINITZ, C. AND P. ROGERS. *A Systems Analysis Model of Urbanization and Change: An Experiment in Interdisciplinary Education*. Cambridge, Mass.: MIT Press, 1970.

287. STENT, GUNTHER S. *The Coming of the Golden Age*. Garden City, NY: The Natural History Press, 1969.

288. STEVENS, PETER S. *Patterns in Nature*. Boston: Little, Brown and Co., 1974.

289. STEWART, JOHN Q. 'Demographic Gravitation: Evidence and Applications,' *Sociometry*, XI, 1948, 31–58.

290. STEWART, JOHN Q. 'The Development of Social Physics,' *American Journal of Physics*, XVIII, 1950, 239–53.

291. STEWART, JOHN Q. 'Empirical Mathematical Rules Concerning the Distribution and Equilibrium of Population,' *Geographical Review*, XXXVII, 1947, 461–85.

292. STEWART, JOHN Q. 'A Measure of the Influence of a Population at a Distance,' *Sociometry*, V, 1942, 63–71.

293. STEWART, JOHN Q. AND WILLIAM WARNTZ. 'Physics of Population Distribution,' *Journal of Regional Science*, I, no. 1, 1958, 99–123.

294. STEWART, JOHN W. 'The Theory of Dimensions and Its Application to the Cataloguing of Physical Quantities,' Unpublished Senior Thesis, Department of Physics, Princeton University, 1949.

295. STODDART, D.R. 'Geography and the Ecological Approach: the Ecosystem as a Geographic Principle and Method,' *Geography*, L, 1965, 242–51.

296. STODDART, D.R. 'Organism and Ecosystem as Geographical Models,' in R.J. Chorley and P. Haggett (eds), *Models in Geography*. London: Methuen, 1967, 511–48.

297. STRAHLER, ARTHUR N. 'Dimensional Analysis Applied to Fluvially Eroded Landforms,' *Bulletin of the Geological Society of America*, LXIX, 1958, 279–300.

298. STRAHLER, ARTHUR N. 'Hypsometric (Area–Altitude) Analysis of Erosional Topography,' *Geological Society of America Bulletin*, LXIII, 1952, 1117–42.

299. STRAHLER, ARTHUR N. *Physical Geography*, third edition. New York: John Wiley, 1969.

300. STROMQUIST, W. 'The Four-Color Theorem for Small Maps,' *Journal of Combinatorial Theory*, XIX B, 1975, 256–68.

301. TAAFFE, EDWARD J. AND HOWARD L. GAUTHIER. *Geography of Transportation*. Englewood Cliffs, NJ: Prentice-Hall, 1973.

302. TAAFFE, E.J., R.L. MORRILL, AND P.R. GOULD. 'Transport Expansion in Underdeveloped Countries: A Comparative Analysis,' *The Geographical Review*, LIII, 1963, 503–29.

303. TAEUBER, KARL AND ALMA TAEUBER. *Negroes in Cities*. Chicago: Aldine, 1965.

304. TAYLOR, PETER J. 'An Interpretation of the Quantification Debate in

British Geography,' *Transactions,* Institute of British Geographers, NS I, 1976, 129–42.
305. TAYLOR, P.J. 'Distances Within Shapes: An Introduction to a Family of Finite Frequency Distributions,' Discussion Paper No. 16, Department of Geography, University of Iowa, 1970.
306. THOM, RENÉ. 'Structuralism and Biology,' in C.H. Waddington (ed.), *Toward a Theoretical Biology,* Vol. 4. Chicago: Aldine, 1972, 68–82.
307. THOM, RENÉ. *Structural Stability and Morphogenesis,* translated by D.H. Fowler. Reading, Mass.: W.A. Benjamin, 1975.
308. THOM, RENÉ. 'Topological Models in Biology,' in C.H. Waddington (ed.), *Towards a Theoretical Biology,* Vol. 3. Chicago: Aldine-Atherton, 1970, 89–116.
309. THOMPSON, D'ARCY. *On Growth and Form,* abridged edition, J.T. Bonner (ed.). London: Cambridge University Press, 1971.
310. TINKLER, KEITH J. 'Bounded Planar Networks: A Theory of Radial Structures,' *Geographical Analysis,* IV, 1972, 5–33.
311. TINKLER, KEITH J. *An Introduction to Graph Theoretical Methods in Geography.* Concepts and Techniques in Modern Geography (CATMOG), paper No. 14. Norwich, UK: Geo Abstracts Ltd, undated.
312. TINKLER, KEITH J. 'On Functional Regions and Indirect Flows,' *Geographical Analysis,* VIII, 1976, 205–13.
313. TINKLER, KEITH J. 'The Topology of Rural Periodic Market Systems,' *Geografiska Annaler,* LV B, 1973, 121–33.
314. TOBLER, WALDO R. 'Geographic Area and Map Transformations,' *Geographical Review,* LIII, 1963, 59–78.
315. TOBLER, WALDO R. 'Map Transformations of Geographic Space,' Unpublished PhD Thesis, Department of Geography, University of Washington, 1961.
316. TOBLER, WALDO R. 'On Geography and Geometry,' Unpublished MS, Department of Geography, University of Michigan, 1966.
317. TOBLER, WALDO R. 'Satellite Confirmation of Settlement Size Coefficients,' *Area,* no. 3, 1969, 30–34.
318. TOBLER, W.R., H.W. MIELKE, AND T.R. DETWYLER, 'Geobotanical Distance Between New Zealand and Neighboring Islands,' *Bioscience,* May 1970, 537–42.
319. TOBLER, W.R. AND S. WINEBERG. 'A Cappadocian Speculation,' *Nature,* 231, May 1971, 39–41.
320. TORNQVIST, GUNNAR, *Studier i Industrilokalisering.* Stockholm: Geografiska Institutionen vid Stockholms Universitet, 1963.
321. TSCHIERSKE, HILMAR. 'Raumfunktionelle Prinzipien in einer allgemeine theoretischen Geographie,' *Erdkunde,* 1961, 92–110.
322. TUAN, YI-FU. *Man and Nature.* Commission on College Geography, Resource Paper No. 10, Association of American Geographers, Washington, DC, 1971.
323. ULLMAN, EDWARD L. 'Geography as Spatial Interaction,' in D. Revzan and E.S. Englebert (eds), *Interregional Linkages.* Berkeley: University of California Press, 1954, 1–12.
324. ULLMAN, EDWARD L. 'The Role of Transportation and the Bases of Interaction," in W. Thomas Jr. (ed.), *Man's Role in Changing the Face of the Earth.* Chicago: University of Chicago Press, 1956.

325. VAN PAASSEN, C. *The Classical Tradition of Geography*. Groningen: J.B. Wolters, 1957.

326. VIDAL DE LA BLACHE, P. *Tableau de la géographie de la France*. Paris: Armand Colin, 1903.

327. VINE, F.J. 'Sea-Floor Spreading,' in I. Gass, P. Smith and R. Wilson (eds), *Understanding the Earth*, second edition. Cambridge, Mass.: MIT Press, 1972, 233–49.

328. WADDINGTON, C.H. 'Form and Information,' in C.H. Waddington (ed.), *Toward a Theoretical Biology*, Vol. 4. Chicago: Aldine-Atherton, 1972, 10–145.

329. WADDINGTON, C.H. 'Thinking About Complex Systems,' *Ekistics*, XXXII, no. 193, 1971, 410–12.

330. WADDINGTON, C.H. (ed.). *Toward a Theoretical Biology*, vols 1–4. Chicago: Aldine, 1968–72.

331. WAGSTAFF, J.M. 'A Possible Interpretation of Settlement Pattern Evolution in Terms of "Catastrophe Theory",' *Transactions*, Institute of British Geographers, NS III, 1978, 165–78.

332. WALMSLEY, D.J. *Systems Theory: A Framework for Human Geographical Enquiry*. Australian National University, Department of Human Geography, publication HG/7, 1972.

333. WARNTZ, WILLIAM. 'Conceptual Breakthroughs in Geography,' *Geographical Perspectives*, no. 33, 1974, 37–40.

334. WARNTZ, WILLIAM. *Distances in a Man-Made Environment*. Unpublished MS.

335. WARNTZ, WILLIAM. 'Global Science and the Tyranny of Space,' *Papers of the Regional Science Association*, XIX, 1967, 7–19.

336. WARNTZ, WILLIAM. *Macrogeography and Income Fronts*. Philadelphia: Regional Science Research Institute, 1965.

337. WARNTZ, WILLIAM. 'New Geography as General Spatial Systems Theory – Old Social Physics Writ Large?' in R. Chorley (ed.), *Directions in Geography*. London: Methuen, 1974, 89–126.

338. WARNTZ, WILLIAM. 'A Note on Surfaces and Paths and Applications to Geographical Problems,' Michigan Inter-University Community of Mathematical Geographers, Discussion Paper No. 6, 1965.

339. WARNTZ, WILLIAM. 'Some Elementary and Literal Notions about Geographical Regionalization and Extended Venn Diagrams,' in John Nystuen (ed.), 'The Philosophy of Maps,' Michigan Inter-University Community of Mathematical Geographers, Discussion Paper No. 12, 1968.

340. WARNTZ, WILLIAM. 'Stream Ordering and Contour Mapping,' *Journal of Hydrology*, XXV, 1975, 209–27.

341. WARNTZ, WILLIAM. 'The Topology of a Socio-Economic Terrain and Spatial Flows,' *Papers of the Regional Science Association*, XVII, 1966, 47–61.

342. WARNTZ, WILLIAM. *Toward a Geography of Price*. Philadelphia: University of Pennsylvania Press, 1959.

343. WARNTZ, WILLIAM. 'Transportation, Social Physics, and the Law of Refraction,' *The Professional Geographer*, IX, no. 4, 1957, 2–7.

344. WARNTZ, WILLIAM. 'The University and the Region – Trends in the Geography of and for the Educational Establishment,' Paper presented at the Northeast Regional Science Association Meetings, Cornell University, April 1976.

345. WARNTZ, WILLIAM AND WILLIAM BUNGE. 'Geography – The Innocent Science.' Unpublished MS.

346. WARNTZ, WILLIAM AND NIGEL WATERS. 'Network Representations of Critical Elements of Pressure Surfaces,' *The Geographical Review*, LXV, 1975, 476–92.

347. WARNTZ, WILLIAM AND MICHAEL WOLDENBERG. 'Concepts and Applications – Spatial Order,' *Harvard Papers in Theoretical Geography*, Paper 1, Harvard University, 1970.

348. WARNTZ, WILLIAM AND PETER WOLFF. *Breakthroughs in Geography*. New York: New American Library, 1971.

349. WATSON, J.W. 'Geography: A Discipline in Distance,' *Scottish Geographical Magazine*, LXXI, 1955, 1–13.

350. WEBB, W.A. 'Analysis of the Martian Canal Network,' *Proceedings of the Astronomical Society of the Pacific*, LXVII, 1955, 283–92.

351. WHITE, ROGER W. 'A Generalization of the Utility Theory Approach to the Problem of Spatial Interaction,' *Geographical Analysis*, VIII, 1976, 39–46.

352. WHITE, ROGER W. 'Sketches of a Dynamic Central Place Theory,' *Economic Geography*, L, 1974, 219–27.

353. WHITEHEAD, ALFRED N. *Process and Reality*. New York: Fordham University Press, 1929.

354. WHITROW, G.J. 'Why Physical Space Has Three Dimensions,' *General Systems*, VII, 1962, 121–9.

355. WHORF, BENJAMIN L. *Language, Thought, and Reality*. New York: The Technology Press and John Wiley, 1956.

356. WIENER, NORBERT. *Cybernetics*. New York: John Wiley, 1948.

357. WILBANKS, THOMAS J. 'Prospects for a General Theory of Space–Time Phenomena,' presented to the special session on 'Theory Synthesis for Periodic Space-Time Phenomena,' Annual Meeting of the Association of American Geographers, Seattle, 1974.

358. WILBANKS, THOMAS J. AND RICHARD SYMANSKI. 'What is Systems Analysis?' *The Professional Geographer*, XX, 1968, 81–5.

359. WILSON, ALAN G. *Entropy in Urban and Regional Modelling*. London: Pion, 1970.

360. WILSON, ALAN G. 'Theoretical Geography: Some Speculations,' *Transactions*, Institute of British Geographers, no. 57, 1972, 31–44.

361. WILSON, EDWARD O. *Sociobiology: The New Synthesis*. Cambridge, Mass.: The Belknap Press of Harvard University Press, 1975.

362. WOLDENBERG, MICHAEL J. 'An Allometric Analysis of Urban Land Use in the United States,' *Ekistics*, 215, 1973, 282–90.

363. WOLDENBERG, MICHAEL J. 'Allometric Growth in Social Systems,' *Harvard Papers in Theoretical Geography*, Paper 6, Harvard University, 1971.

364. WOLDENBERG, MICHAEL J. 'The Hexagon as a Spatial Average,' *Harvard Papers in Theoretical Geography*, Paper 42, Harvard University, 1970.

365. WOLDENBERG, MICHAEL. 'Hierarchical Systems: Cities, Rivers, Alpine Glaciers, Bovine Livers, and Trees,' *Harvard Papers in Theoretical Geography*, Paper 19, Harvard University, 1968.

366. WOLDENBERG, MICHAEL J. 'Relations Between Horton's Laws and Hydraulic Geometry as Applied to Tidal Networks,' *Harvard Papers in Theoretical Geography*, Paper 45, Harvard University, 1972.

367. WOLDENBERG, MICHAEL J. 'Spatial Order in Fluvial Systems: Horton's Laws Derived from Mixed Hexagonal Hierarchies of Drainage Basin Areas,' *Harvard Papers in Theoretical Geography*, Paper 13, Harvard University, 1968.

368. WOLDENBERG, MICHAEL J. 'A Structural Taxonomy of Spatial Hierarchies,' *Harvard Papers in Theoretical Geography*, Paper 39, Harvard University, 1970.

369. WOLDENBERG, MICHAEL AND BRIAN J.L. BERRY. 'Rivers and Central Places: Analogous Systems?' *Journal of Regional Science*, VII, no. 2, 1967, 129–39.

370. WOLDENBERG, MICHAEL, GORDON CUMMING, KEITH HARDING, KEITH HORSFIELD, KEITH PROWSE, AND SHIAM SINGHAL. 'Law and Order in the Human Lung,' *Harvard Papers in Theoretical Geography*, Paper 41, Harvard University, 1970.

371. WOLFF, ROBERT PAUL. *The Ideal of the University*. Boston: Beacon Press, 1969.

372. WOLPERT, LEWIS. 'The Concept of Positional Information and Pattern Formation,' in C.H. Waddington (ed.), *Toward a Theoretical Biology*, Vol. 4. Chicago: Aldine, 1972, 83–94.

373. WOLPERT, LEWIS. 'Pattern Formation in Biological Development,' *Scientific American*, CCXXXIX, October 1978, 154–64.

374. WRIGLEY, E.A. 'Changes in the Philosophy of Geography,' in R. Chorley and P. Haggett (eds), *Frontiers in Geographical Teaching*. London: Methuen, 1970, 3–20.

375. YEATES, MAURICE H. 'Hinterland Delimitation: a Distance Minimizing Approach,' *The Professional Geographer*, XV, 1963, 7–10.

376. YOUNG, E.C. *The Movement of Farm Population*. Ithaca: Cornell Agricultural Experiment Station, Bulletin 426, 1924.

377. ZEEMAN, E.C. 'Catastrophe Theory,' *Scientific American*, CCXXXIV, no. 4, April 1976, 65–83.

378. ZIPF, GEORGE K. *National Unity and Disunity*. Bloomington, Ind.: University of Indiana Press, 1941.

379. ZIPF, G.K. 'The P_1P_2/D Hypothesis: On the Intercity Movement of Persons,' *American Sociological Review*, XI, 1946, 677–86.

380. ZOBLER, L. 'Decision Making in Regional Construction,' *Annals,* Association of American Geographers, XLVII, 1958, 140–8.

Name Index

Subject Index

E DUF